# Malaysian water sector reform

# Malaysian water sector reform

## Policy and performance

### Ching Thoo Kim

*Environmental Policy Series – Volume 8*

**Wageningen Academic**
**P u b l i s h e r s**

Wageningen Academic Publishers
P.O. Box 220
6700 AE Wageningen
The Netherlands
www.WageningenAcademic.com
copyright@WageningenAcademic.com

ISBN: 978-90-8686-219-1
e-ISBN: 978-90-8686-773-8
DOI: 10.3920/978-90-8686-773-8

First published, 2012

© Wageningen Academic Publishers
The Netherlands, 2012

# Preface

'There is no life without water. We do not inherit water from our grandfathers, but instead borrow it from our grandchildren. There is no alternative to water'. We often hear water managers and policy makers chanting these mantras. Yet millions of people on this planet are still without adequate access to clean and drinkable water. Technological solutions and economic instruments have so far been unable to provide a solution to tackling water problems. On the contrary, solving water issues always involves (difficult) political considerations as well as (rigid) institutional reforms that need to balance the socio-political (including environmental) interests of the state, the economic interests of market actors and the interests of consumers, civil society and non-governmental organizations. This research focuses on the reform of the Malaysian water sector. It seeks to explain the policy perspectives that underlay the reform process and to examine the extent in which the outputs of the reform have contributed in attaining the reform's objectives. It also analyses the impacts of the reform in improving the operational efficiency and environmental performance of water utilities.

This thesis is the result of four and half years of research work resulting from my involvement (since 2004) in the reform of the Malaysian water sector. My passion for water grew even stronger when, in 2008, I decided to translate it into a pursuit of academic excellent by enrolling into a PhD programme at Wageningen University. This endeavour would not have been possible without financial supports from my employee – the Government of Malaysia – for which I am truly indebted.

I thank Prof. Wim van Vierssen, the Director of KIWA Water Research (now KWR Watercycle Research Institute) for connecting me to the university (and the Environmental Policy Group – ENP). He shows great interest in my research ideas during our meeting in Singapore in 2007. I humbly apologize for not being able to consider him as one of my supervisors.

At ENP, I had the privilege to work under the supervision of Prof. Arthur Mol. I greatly benefitted from his critical and analytical approaches that contributing greatly to the successful completion of this thesis. My day to day supervisor was Dr. Bas van Vliet who has been excellent in providing clear directions and feedback throughout the entire period of this study. We share a common understanding and knowledge of the water sector which helped to facilitate smooth and effective communications. I would like to record my deepest appreciation to both of them for the valuable contributions they made to this study. I enjoyed engaging in delightful deliberations with the rest of the ENP staff which often, directly or indirectly, contributed to this study. Back home in Malaysia, I am grateful to my local supervisor, Prof. Chan Ngai Weng of the Universiti Sains Malaysia for his thoughts that sharpened and fine-tuned the early version of my proposal.

I thank Corry Rothuizen for her administrative assistance and support, both prior to my arrival in the Netherlands and during the course of this study. The international setting of ENP has made my four and a half years stay here an interesting and memorable one. I enjoyed the company of Sammy Letema, Leah Ombis, Laurent Glin, Kanang Kantamaturapoj, Kim Dung, Carolina Maciel, Elizabeth Sargant, Liu Wenling, Radhika Borde, Tung Son Than, Natapol Thongplew, Joeri Naus, Debasish Kundu, Marjanneke Vijge, Javed Ali Haider, Hilde Toonen, Dorien Korbee, Sarah Stattman, Jennifer Lenhart, Alice Miller, Alexey Pristupa, Li Jia, Zheng Chaohui, Feng

Yan, Belay Mengistie, Tran Thi Thu Ha, Jin Shuqin and Somjai Nupueng. I feel honoured to have shared the same working room with Eira Bieleveldat-Carballo Cardenas, Judith Floor and Harry Barnes-Dabban, whom (coincidently or not) happen to come from (four) different continents: North America, Europe, Africa and Asia. A truly global workplace!

In Wageningen I also met Phi Jane and Marcel whom I consider to be like my own family away from home. On behalf of my family, I want to express my sincere appreciation to both of them for warmly accepting us and for the delicious (Thai) foods you served us.

Wageningen only has a relatively 'small' Malaysian community – but we are always 'big' when it comes to food. I thank all Warga-Wage for all the great and fantastic 'makan-makan' sessions. I cannot go without recording my since appreciation to Shahrul Ismail and family, Maimunah Sany and family, Azlim Khan and family, Razak and family, Naim and family, Norhariani and family, Asyraf and family, Jimmy and family, Izan and family, Nurulhuda Khairudin, Norulhuda Ramli, Freddy Yeo, Noor Liyana, Loo Wee and Rui Jack Chong. A special appreciation is dedicated to 'Uncle' Alan Sim and his wife.

This study would have not been possible without the cooperation from those who responded to surveys, gave up time for interviews, etc. I am deeply touched by the assistance provided to me: water operators; government officials; consumer associations; and environmental organizations and would like to thank them all. Unfortunately, I am not able to list every name on this page (but they are all in the Appendix). In addition, my special appreciation goes to Ir. Jaseni Maidinsa, General Manager of PBAPP and Dato' Adzmi Din, Chief Executive Officer of SADA for allowing me to conduct the case studies.

I should also like to thank Nick Parrott of TextualHealing.nl for the English editing and helped me produce this thesis.

Last but not least, my deepest gratitude goes to my mum, dad and in-laws for their persistent love and support throughout my life. I thank my wife Duangrat for the sacrifices and suffering that she endured in nurturing our children while I was away. I owe my children Adeesakh and Praphaicit a lot. They provided me with the strength and inspiration to complete this study.

Wageningen/Putrajaya
October 2012

# Table of contents

# List of abbreviations

| | |
|---|---|
| ABASS | Konsortium ABASS |
| AGC | Attorney General's Chamber |
| AKSB | Air Kelantan Sendirian Berhad |
| AUIB | Air Utara Indah Berhad |
| AWER | Association of Water and Energy Research Malaysia |
| AWWA | American Water Works Association |
| AWWARF | American Water Works Association Research Foundation |
| BAKAJ | Badan Kawalselia Air Johor |
| BAKAS | Badan Kawalselia Air Selangor |
| BOOT | build, operate, own, transfer |
| BOT | build, operate, transfer |
| CAP | Consumers' Association of Penang |
| CAWP | Coalition Against Water Privatization |
| DMZ | district metering zone |
| DOE | Department of Environment |
| EEA | European Environmental Agency |
| EQA | Environmental Quality Act |
| ESB | Equiventures Sendirian Berhad |
| FOMCA | Federation of Malaysian Consumers Associations |
| LAP | Lembaga Air Perak |
| LUAN | Lembaga Urus Air Kedah |
| MEWC | Ministry of Energy, Water and Communications |
| MLD | million litres per day |
| MNC | multinational corporations |
| MTUC | Malaysia Trade Union Congress |
| MUC | Metropolitan Utilities Corporation |
| MWA | Malaysia Water Association |
| MWIG | Malaysia Water Industry Guide |
| NGDWS | National Guideline for Drinking Water Standard |
| NGOs | non-governmental organisations |
| NRW | non-revenue water |
| NWRC | National Water Resources Council |
| NWSC | National Water Services Commission |
| OFWAT | Office of Water |
| PAAB | Pengurusan Aset Air Berhad |
| PAM | Perbadanan Air Melaka |
| PBAHB | Perbadanan Bekalan Air Holdings Berhad |
| PBAPP | Perbadanan Bekalan Air Pulau Pinang |
| PCPA | Penang Consumer Protection Association |
| PNSB | Puncak Niaga Sendirian Berhad |

| PPP | public-private partnership |
| PSP | private sector participation |
| PUAS | Perbadanan Urus Air Selangor Berhad |
| PWD | Public Works Department |
| SADA | Syarikat Air Darul Aman |
| SAINS | Syarikat Air Negeri Sembilan |
| SAJH | Syarikat Air Johor Holdings |
| SAM | Sahabat Alam Malaysia |
| SAMB | Syarikat Air Melaka Berhad |
| SAP | Syarikat Air Perlis Sdn Bhd |
| SATU | Syarikat Air Terengganu |
| SPANA | Suruhanjaya Perkhidmatan Air Negara Act |
| SWC | Southern Water Corporation |
| Syabas | Syarikat Bekalan Air Selangor Sdn. Bhd. |
| TNB | Tenaga Nasional Berhad |
| TWSB | Taliworks Sendirian Berhad |
| UNDP | United Nations Development Programme |
| UNICEF | United Nations International Children's Emergency Fund |
| WHO | World Health Organisation |
| WSIA | Water Services Industry Act |
| WTPs | water treatment plants |
| YTL | Yeoh Tiong Lay |

# Chapter 1.
# Introduction

## 1.1 The historical development of the water supply sector in Malaysia

This research focuses on the water sector in Malaysia. The problems in the Malaysian water sector can be traced back to the historical development of the water sector from the 19th century onwards. During that period, the British colonial administration established a water supply system in the Federated Malay States and the Straits Settlements (Jabatan Bekalan Air, 2012). The first water supply was piped into Penang in 1804, followed by Kuching, the capital city of Sarawak in 1887. Four years later, the capital city of Kuala Lumpur had its water supply system, and then Melaka in 1889 (Jabatan Bekalan Air, 2012). By 1939, households in the major towns in Malaya (as it was then known) were served with piped-water. However, during the Japanese Occupation (1941-45), most of the water installations deteriorated. By 1950, the water installations had been restored and water supply reached more than 1.15 million people. The demand for piped water continued to grow steadily after Malaya gained independence from British in 1957. In 1959, the Klang Gates Dam and the Bukit Nanas treatment plant were built, to meet the demand from Kuala Lumpur and the surrounding areas (Jabatan Bekalan Air, 2012).

After independence, the Water Supply Branch of the federal Public Works Department (PWD) provided consultation and technical advice to state water supply authorities, and coordinated all the water supply projects that were funded by federal loans and grants. At this time, in several states (i.e. Perlis, Kedah, Labuan and Sarawak) the state PWD administered water supply, roads, buildings and mechanical and electrical works. As time went by, state governments (Negeri Sembilan, Pahang and several others) took over responsibility for the water supply, often creating state water departments. They were often assisted in this task by engineers seconded from the federal PWD. In Kelantan and Terengganu state water departments were established in a bid to boost efficiency and effectiveness. These water corporations embraced a private sector working culture such as commercial accounting and the use of performance indicators, but they remained public entities and subsidiaries of the state governments. Meanwhile, a hybrid model, in the form of state water boards, emerged in the states of Perak, Melaka and (in several districts of) Sarawak. In order to keep up with the rapid economic growth of the late 1980s and the 1990s, state governments turned to private operators to fund the further development and expansion of water infrastructure. The states of Selangor, Johor, Kelantan[1] and Penang decided to privatize their entire water supply system, while the states of Kedah, Perak and Sabah privatized parts of their water services such as the operation and maintenance of water treatment plants (MEWC, 2008). Figure 1.1 summarizes the types of water supply entities in Malaysia in 2006, prior to the implementation of the water sector reform.

The development of water infrastructure was funded by the federal government under 5-year development programmes, known as the Malaysia Plans. The Economic Planning Unit records show that the first allocation for the water sector was made under the Third Malaysia Plan (1976-

---

[1] Later the state government bought the privatized entity bringing it back under state control and ownership.

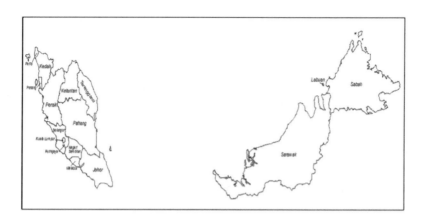

| State | Water entities | Type of operations |
|---|---|---|
| Perlis | PWD Perlis | public water unit |
| Kedah | PWD Kedah | public water unit |
| | AIUB | private (treatment only) |
| | Taliworks | private (treatment & distribution in Langkawi Island only) |
| Penang | PBAPP | private |
| Perak | LAP | water board |
| | MUC | private (treatment only) |
| | GSL Water | private (treatment only) |
| Selangor (incl. Kuala Lumpur and Putrajaya) | PNSB | private (treatment only) |
| | Abass | private (treatment only) |
| | Splash/Gamuda Water | private (treatment only) |
| | Syabas | private (distribution only) |
| Negeri Sembilan | JBA NS | state water department |
| | Salcon | private (treatment only) |
| Melaka | PAM | water board |
| Johor | SWC | private (treatment only) |
| | ESB | private (treatment only) |
| | SAJH | private (distribution only) |
| Pahang | JBA Pahang | water department |
| Terengganu | SATU | water corporation |
| Kelantan | AKSB | water corporation |
| Sabah | Sabah Water Department | state water department |
| | Timatch Water | private (treatment only) |
| | Jetama Sdn Bhd | private (treatment only) |
| Sarawak | Sarawak Public Works | public water unit |
| | Sibu Water Board | water board |
| | Miri Water Board | water board |
| Labuan | JBA Labuan | federal water department |
| | Encorp Utilities | private (treatment operator for Beaufort treatment plant) |

*Figure 1.1. List of water entities in Malaysia in 2006 (MWA, 2006).*

1980) (Economic Planning Unit, 2008). During that period, a total of RM 538 million (approx. US$ 178 million)[2] was set aside for the water sector with the main focus on developing rural water supplies. The rapid economic growth[3] experienced during this period allowed the federal government to allocate large investments for the water sector in the subsequent Malaysia Plans (Economic Planning Unit, 2010). From RM 2.1 billion allocated in the Fourth Malaysia Plan (1981-1985), the allocations went up to RM 2.3 billion under the Fifth Malaysia Plan (1986-1990) and then declined slightly to RM 2.1 billion in the Sixth Malaysia Plan (1991-1995). Under the Eighth Malaysia Plan (2001-2005), investments in the sector doubled to over RM 4 billion, and more than doubled again (to RM 9.8 billion) under the Ninth Malaysia Plan (2006-2010). Prior to the water sector reform (1997-2006), Malaysia was regarded as one of the Asian countries with the highest access rate to water supply and adequate sanitation, as shown in Figure 1.2 (Economic Planning Unit, 2005; World Bank, 2012). The Asian Development Bank (2006) rated Malaysia as 'on track' in meeting the Target 10 of the United Nations' Millennium Development Goal by 2015.

More recently the World Bank (2012) reported that Malaysia had made good progress in providing clear water and adequate sanitation to its citizens over the past decade and access to water and sanitation was above regional and world averages (Figure 1.3).

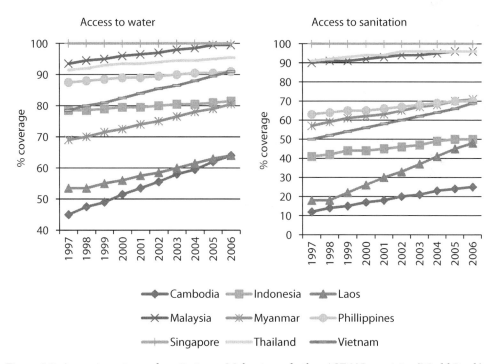

*Figure 1.2. Access to water and sanitation – Malaysia and other ASEAN countries (World Bank).*

---

[2] At an exchange rate of US$ 1 = RM 3.02 on 1 May 2012. See http://www.oanda.com. This rate is applied throughout the thesis.

[3] From 1991-2010 GDP grew at an average of 5.8%.

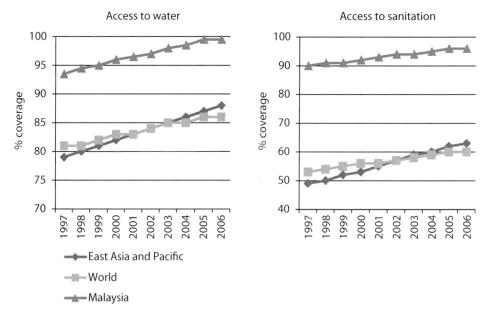

*Figure 1.3. Access to water and sanitation – Malaysia, East Asia and the Pacific and the World (World Bank).*

## 1.2 The problems facing the Malaysian water sector prior to reform

Prior to reform water provisioning in Malaysia was predominantly the responsibility of public authorities – federal government provided the financial assistance and the state governments managed water supply and regulation. Generally, water was provided by three categories of service providers: state water departments; state water corporations; and private operators. Despite the improvements in the sector and major achievements in terms of access and coverage, at the start of the new Millennium the Malaysian water sector still faced many problems, which can be summarized as falling into four main categories: (1) operational inefficiency; (2) ineffective governance and regulation; (3) budgetary constraints; and (4) poor environmental performance.

The operational inefficiency in the water sector was caused by two main factors: the inability of water utilities to reduce the high volume of non-revenue water and the existence of a below-cost tariff structure. Despite making some improvements, some states were still registering losses as high as 50%, an indication of serious operational problems. This problem seriously undermined their ability to generate enough revenues to sustain their operations and to expand services to new (especially rural) areas. In some states, water access – such as Kelantan, Sabah and Sarawak – was quite low. Among these three states, Kelantan recorded the lowest coverage of 64%, 65% and 72% in the years 2004-2006 (MWA, 2005, 2006, 2007). Almost 50% of water users in Kelantan relied on well water (AWER, 2011).

In addition, the presence of a below-cost tariff structure further aggravated the problems of inadequate revenue. Water utilities were unable to cover the costs – for abstraction, purification, transportation – incurred in providing water to users. Political reasons (fear of losing votes) and socio-economic reasons (promoting equity and accessibility) inhibited state governments from implementing a full cost recovery model in 75% of public water departments (MWA, 2005). In some states (i.e. Labuan and Sabah) the prevailing tariffs had been in place since 1982 without any revision. In 2007, more than half of the country's water utilities recorded a deficit on their balance sheet as a result of below-cost tariff structures (MWA, 2008). As a result, water utilities did not generate sufficient revenue to expand their networks to new areas or even maintain the existing ones.

The policy of subsidizing domestic water tariffs in Malaysia also led Malaysians to consume more water than other comparable countries in the region, and proved to be harmful to the environment. For instance, in 2002 every Malaysian household consumed an average of 283 litres/capita/day of water, far more than the 165 litres recommended by the United Nations (MWA, 2004). During the same period, water-stressed Singapore only consumed 165 litres/capita/day and was making plans to reduce this to 147 litres by 2020. Households in the Philippines and China consume 86 and 164 litres/capita/day respectively (PUB, 2010; Asian Development Bank, 2010; UNDP, 2006). In some Asian metropolitan areas water consumption levels are especially high: Bangkok and Tokyo record levels of 430 and 374 litres/capita/day respectively (UNDP, 2006). High water consumption levels create pressure on the available water sources and supply infrastructure, and can degrade the environment through waste water discharge, lower groundwater tables and the need to clear land to build water infrastructure (Asian Development Bank, 2007).

This policy of subsidizing water tariffs contradicts the concept of using pricing mechanisms to promote the efficient use of water. Many countries, including Malaysia, have not been successful in implementing a full cost recovery tariff structure, thus leading consumers to use water inefficiently (Zetland, 2011). This policy also led to Malaysia having some of the lowest water tariffs in the world (Table 1.1), a policy that did promote high levels of access to water, even among the poorer sections of society (Biswas & Tortajada, 2010). Some states, such Penang and Terengganu, have water tariffs that are far below the national average (MWA, 2005).

*Table 1.1. Average water tariff in Malaysia and selected Asian countries (2006) (World Bank, 2008; Asian Development Bank, 2008; MWA, 2006).*

| Countries | Average water tariff (RM sen/m³) | Remarks |
|---|---|---|
| Malaysia | 0.97 | - |
| Vietnam | 0.89 | Ho Chi Minh City |
| Thailand | 1.19 | Under the service area of Metropolitan Waterworks Authority |
| China | 1.86 | Shenzhen province (above 31 m³ effective from 2004-2011) |
| Philippines | 2.39 | Under the service area of Maynilad Water Services concession in Manila (up to 40 m³) |
| Singapore | 2.84 | Effective from July 2000 |

In summary, at the start of the new Millennium the Malaysian water sector was in urgent need of reform to remove the main causes of high water loss and low tariffs.

The water sector also lacked an effective governance system, as there was a lack of clear separation of power between the tasks of formulating policy, regulating water provisioning and supplying water. All three functions were vested under the sole jurisdiction of state governments, the rightful owner of water sources (as specified under Schedule Nine of the Federal Constitution). This constitutional right of state governments limited the extent to which the Federal government could interfere in this matter, if for example states ruled in favor of particular rather than public interests. This complicated involvement of state governments in water businesses (through business partnerships with private operators) raised several concerns. First, it sometimes gave rise to conflicts of interests as the party entrusted to regulate water provisioning (state governments) was also participating in the water business (MEWC, 2008). Second, the state government's role as protectors of citizen's right to water was questioned as they were not able to act independently (as they were influenced by business decisions of private parties). In Selangor and Johor (where state governments were involved in water business) state governments had little influence on decisions about tariffs as private operators had exclusive rights over tariffs. This prompted civil society groups, such as Water Watch Penang to equate private involvement in the water sector to an act of 'piratization': since private water operators were allowed to make profits by using public money and at the expense of water users at large.

Thirdly, limited public funding (in addition to high water losses and low tariffs) continued to contribute to the problem of under-investment in water infrastructure. The budgets allocated under the five-year Malaysia Plans fell short of what was needed by state governments to maintain and expand water infrastructure, and were also under competition from other sectors. For instance, only half of the (RM 22 billion) required budget for water infrastructure for 2000-2010 was allocated to the sector (Economic Planning Unit, 2008). This resulted in delays in investment in several crucial areas, such as tackling water loss, improving water supply coverage and modernizing information and IT systems. The majority (if not all) of state governments began to turn to (richer) private water operators for financial assistance and entered into various forms of public-private partnerships, ranging from (short-term) service contracts to (long-term) concessions. However, private sector involvement in the water sector created several problems. First, fragmented privatization – where profitable (treatment) operation was given to private operators along with exclusive rights for tariff revision – did not generate any extra revenue for state governments. Sometimes the opposite: state governments had to shoulder heavier financially burdens (from compensation) when they tried to prevent tariff increases. Second, the state regulatory bodies were weak and relatively ineffective in enforcing the conditions of the concession contracts (MEWC, 2008). Not only they were inexperienced in effectively using regulatory tools – performance indicators, benchmarking, price capping mechanism, etc., but they were also subject to the rules and discretions of the state administration and were often ineffective in dealing with private operators that usually had political and business connections with the state administration.

Lastly, the water sector urgently needed to address the problems of poor environmental performance related to sludge management, information disclosure and – to a lesser extent – water quality. While water utilities' compliance with water quality standards, as prescribed in the National Guideline for Drinking Water Quality Standard 2001, was often satisfactory, sludge management and environmental information disclosure were definitely not.

The growing demand for water supply – from 7,108 million litres per day (MLD) in 2004 to 7,628 MLD in 2006 – forced water utilities to produce more water which eventually led to an increase in the amount of water treatment sludge being produced (MWA, 2004, 2006). Approximately 600 tonnes of sludge were produced daily by 29 water treatment plants managed by Puncak Niaga Sendirian Berhad in Selangor alone, while the Langat 2 Treatment Plant (under construction) is expected to produce another 400-500 tonnes of sludge daily in 2015 (PAAB, 2009). The presence of toxic materials such as aluminium and arsenic requires sludge to be properly treated and disposed of (Makris & O'Connor, 2007). The presence of organic pollutants in Malaysia's rivers increased from 187,555 kg/day in 2004 to 208,441 in 2005 (World Bank, 2012). Most water utilities do not have the sludge treatment facilities (at least sludge lagoons) and as a result directly discharge their sludge into the environment. This damages the environment and threatens the future availability of raw water sources. Water utilities had (and have) little interest in sludge recycling and reuse as there are no incentives from the government and no commercial demand for the materials.

Acquiring information about the Malaysian water sector posed a considerable challenge and much of the data that was available suffered from validity problems (Secretary-General of the Malaysia Water Association, personal interview). Several factors contributed to this: the absence of a clear governmental policy relating to information management; the absence of a single body entrusted to coordinate information management; and funding shortages among public water which prevented them from installing information management systems. Moreover, public disclosure of information was prevented by laws (i.e. The Official Secrets Act 1972 and Banking and Financial Institutions Act 1989). The Centre for Independent Journalism Malaysia (2007) reported that (environmental) information on water privatization, river pollution and drinking water quality is kept from the public. Most of information in the water sector was in the possession of the private water enterprises. These information problems prevented the regulator from fulfilling its regulatory function.

These four problems of the water sector are not unique to Malaysia and can be found in many developing and emerging economies. Often they serve as a catalyst for reforming the water sector (Casarin, Delfino & Delfino, 2007; Hall & Lobina, 2006). This was also the case in Malaysia where, in 2004, the federal government initiated a reform of the water sector with the aim of improving the long term operational efficiency and environmental effectiveness of the water sector. This reform involved several fundamental changes to the legal and regulatory framework – the introduction of Water Services Industry Act (WSIA) and Suruhanjaya Perkhidmatan Air Negara Act (SPANA) – and far sweeping reforms to the institutional structure that saw the creation of a central regulatory authority, state water companies, state water resource regulators and the sector financier. This study attempts to describe and assess the process and the effects of the reform.

## 1.3 Research objectives and questions

The objectives of this research are to understand and explain the policy process of the water sector reform, to examine the extent to which the outputs of the reform met their objectives, and to assess the impacts of the reform on the performance of water utilities in terms of operational efficiency and environmental effectiveness.

The research is guided by three questions:

- How can we understand and explain the policy process of the water sector reform?
- To what extent have the outputs of the reform contributed to the realization of the reform's objectives?
- To what extent has the water sector reform improved the operational efficiency and environmental effectiveness of water utilities?

## 1.4 Research methodology

In this section, the research design is first introduced, followed by a brief explanation about data collection and data analysis methods and the role of the researcher in this study. Details of each research method are presented in the relevant chapters.

### 1.4.1 Research design

Kumar (2005) stresses the importance of good research design, which determines how a research is to be conducted. It provides a 'blueprint' which helps the researcher to deal with four problems: what questions to answer, what data are relevant, how data need to be collected and how the results are to be analysed. A good research design helps the researcher to build a 'logical sequence that connects empirical data to study's initial research questions and ultimately, to its conclusions' (Yin, 2009: 26). Generally, there are several choices of research strategy – such as case studies, surveys, experiments, analysis of archival records – and selection of one (or more) will depend on the type of the research questions posed, the extent of the researcher's control over actual events and the degree of focus on contemporary (as opposed to historical) events (Yin, 1994).

The objectives of this research are to understand and explain the events and the policy process of the reform; to identify the outputs of the reform and how they have contributed to reaching the reform's objectives; and to assess the impact of the reform on water operators. In general, this means that this research is focused on answering 'what' and 'how' questions.

The 'what' questions investigate the influence of the reform on operational efficiency and environmental effectiveness. The influence of the reform is evaluated by employing seven indicators – four assigned to operational efficiency, and three to environmental effectiveness.

The 'how' questions focus on understanding and explaining the dynamics of the policy process of the reform and how the outputs of the reform have contributed towards realizing the reform's objectives. Here, more use was made of in-depth interviews, formal and grey literature review and participatory observation (as the author has worked for many years in the water sector).

Besides exploring the effects of the reform on all the water operators, this research includes two comparative case studies – of one public (SADA) and one private (PBAPP) water operator. This provides a more detailed explanation of the extent to which the reform contributed to improving the operational efficiency and environmental effectiveness of water utilities (Figure 1.4). There are several reasons for choosing these two case studies. First, they represent the public and private water utilities (operating along the whole spectrum of water cycle) and show interesting disparities

*Figure 1.4. A map of Peninsular Malaysia indicating case study sites (black – SADA; grey – PBAPP).*

in terms of efficiency and operational effectiveness. Second, by 2010 both companies were fully regulated by the new central regulator – the National Water Services Commission (NWSC) – and subscribed to the financial arrangements established by the sector's financier – the Pengurusan Aset Air Berhad (PAAB). This meant that we could expect to observe the impacts of the reform on these two operators. Third and last, the researcher had good access to (information) sources in these two entities.

### 1.4.2 Methods of data collection

This research combines quantitative and qualitative methods. An advantage of combining different methods, is that it allows for triangulation thus increasing the reliability and validity of the research (Niehof, 1999; Denzin, 1989; Meijer, Verloop & Beijaard, 2002; Modell, 2005). Yin (2003) argues that triangulation can corroborate the data obtained from different sources. This research gathered data from different sources including documents and publications, data sets of water utilities, resource persons in the field, and observations. Various data collection methods were used (Table 1.2).

#### In-depth interviews

A total of 53 in-depth semi-structured interviews (guided by list of questions – see Appendix 1) were conducted with four types of stakeholders – water operators, government officials, consumer

*Table 1.2. Data collection methods.*

| Data collection methods | Tools | Data sources |
|---|---|---|
| In-depth semi-structured interviews | face-to-face interviews | 38 water utility managers<br>7 government officials<br>4 representatives from consumer associations<br>4 representatives from environmental organizations |
| Observations | site visits | 7 water treatment plants<br>5 sludge treatment facilities<br>4 drinking water labs<br>4 customer service centres |
| Secondary data collection | water utilities and governmental data bases, reports, websites, journal articles, books | policy documents, statistical data, subsidiary legislations (rules, regulations), annual reports, publications, consultant's reports, Cabinet papers |
| Surveys | questionnaires | 20 (public, corporatized, private) water utilities<br>35 questionnaires (2 to 'distribution' only, 8 to 'treatment' only and 25 to 'treatment and distribution' water utilities) |

associations and environmental organizations. The interview questions were structured into three parts. The first part sought information regarding respondents' overall perceptions about the reform. In the second part, respondents were asked about the role of the regulator and for their evaluation of WSIA and SPANA, the main legal documents supporting the reform. The third part asked questions related to the implementation of a 'green tax' and other policy interventions that might be used to strengthen the reform. The in-depth interviews were used to understand respondents' points of view, perceptions and their involvement in the reform process, as well as to gather factual information about the reform (Kvale, 1996 in Hemming, 2008).

## Observation

Direct observations, with the aid of checklists, were administered when studying the two case studies. The researcher visited water treatment plants to observe how water sludge was handled and disposed of; drinking water labs for a close look at how these operated and their capabilities and; customer service centres to witness how customers' complaints were handled and managed. Field notes were made during the visits, and were processed directly afterwards. In addition to field notes, relevant evidence from site visits was photographed, as shown in Chapter 6. This method allows the researcher to perform cross-data validity checks on data and the information obtained from the surveys and interviews (Kumar, 2005; Patton; 1999).

*Secondary data collection*

Through this method, extensive reviews of legal documents were performed to determine the key policies related to the water sector reform. This included the WSIA, SPANA, the Environmental Quality Act, the National Guideline for Drinking Water Quality Standards and various reports submitted by consultants on the reform. In addition, reviews of reports (published and unpublished, technical and non-technical), and other related documents were performed to gauge the performance of water operators prior to and after the reform. Included in this review was the Water Industry Guide – between the years 2004 and 2011 – the annual publication of water operators' performance indicators published by the Malaysia Water Association, and utilities' annual reports.

*Surveys*

A total of 35 questionnaires were administered with 20 (public and private) water utilities, including the two case studies: two of these questionnaires were sent to 'distribution only' water utilities, eight to 'treatment only' water utilities and the remaining 25 to 'treatment and distribution' water utilities – as shown in Table 1.2. Only the Peninsular Malaysia-based water utilities (public and private) were surveyed (as those on Sabah and Sarawak were excluded from the reform). This method focused on acquiring quantitative data and information regarding the operational efficiency (Chapter 5) and environmental performance of water utilities (Chapter 6). In both chapters, a detailed description of data collection methods is presented.

### 1.4.3 Methods of data analysis

Data collected from the interviews were tape-recorded. The tape-recorded data were transcribed and classified before analysis. Data were analysed using content analysis, a method to analyse key ideas, phrases and meanings within answers given to interview questions (Weber, 1990). The same method was also used to analyse data obtained from secondary sources. Wherever possible the data is presented quantitatively.

### 1.4.4. Position of the researcher

As highlighted in section 1.4.1, the researcher worked for several years (from 2004-2008) in the Malaysian water sector as an officer within the Ministry of Energy, Water and Communications. During this time the researcher directly participated in the reform process. This direct involvement in the water reform had both advantages and disadvantages.

The advantages included the researcher having wide access to rich sources of information and to key actors in the water sector. By contrast, the participation of the researcher in the reform (having a dual role of investigator and object of research at the same time) could give rise to problems of objectivity and bias in this research. Efforts were taken to minimize such influences on the research. These included diversifying the sources of information (to verify the researcher's ideas through triangulation via multiple interviews, secondary data sources, observations, etc.) and engaging in frank discussions during in-depth interviews to try-out contrasting views and perspectives. The researcher constantly sought to take a 'neutral' position rather than that of a 'government official'. The funding of this research (from the Malaysian government) as well as the academic location where it was carried out (an independent foreign university) prevented stakeholders having an influence on the focus, research design, analysis, conclusion and dissemination of the research.

## 1.5 Organization of the thesis

This thesis consists of seven chapters. Chapter 2 explains the theoretical frameworks used in this research: the policy arrangement and policy evaluation frameworks. The last part of this chapter conceptualizes these theoretical frameworks with respect to the three questions that guided this research.

The empirical material starts with Chapter 3. This chapter describes the events and driving forces behind the water sector reform, identifies and defines the policy objectives and the outputs associated with it, and examines how these outputs have contributed to the attainment of the reform's objectives.

Chapter 4 explains the policy process of the water sector reform, through the prism of the policy arrangement approach. It provides an in-depth analysis of the main discourses in the water sector, the close interactions between state and non-state actors, the resources-power nexus, and the formal and informal rules which structured the reform and which emerged from it.

Chapters 5 and 6 analyse the impacts of the reform on the operational efficiency and environmental effectiveness of water utilities. The analysis uses seven indicators – four relating to operational efficiency and three to environmental effectiveness. Chapter 5 provides a detailed assessment on indicators for non-revenue water, collection efficiency, unit production cost and customer complaints, prior to and after the reform. The same assessment is repeated on sludge management, compliance with drinking water standards and information disclosure in Chapter 6. The two in-depth case studies, comparing public and private operators – SADA and PBAPP – investigate these indicators more thoroughly.

In the concluding Chapter 7, the empirical findings are synthesized to answer the research questions. This chapter also reflects on the theories and research methodology used in the research, the relevance of the research to the existing literature and ends with suggestions for future research and policy recommendations.

# Chapter 2.
# Analysing water supply sector reform: a theoretical framework

## 2.1 Introduction

This chapter firstly presents a general overview of the water supply reform and in so doing explains how several key concepts (equilibrium levels, efficiency, effectiveness, equity, competition and unbundling) are relevant to my analysis of water supply reform in Malaysia. Section 2.3 presents the policy arrangement approach, the theoretical framework used for understanding and analysing the reform process. Section 2.4, discusses the policy evaluation approach, the framework used to measure the outcomes of the reform. This approach draws upon the European Environment Agency's policy evaluation model, the American Water Works Association's QualServe Business Model, and performance indicators set by the Malaysian Water Association. The last section formulates a theoretical framework for this research and explains how this framework helps to answer the research questions.

## 2.2 Water supply reform: an overview

### 2.2.1 The need for reform

This research examines water supply reform in Malaysia, which was initiated and implemented as a public policy initiative by the federal state. Among others, the state's goals for the reforms were to overhaul 'failed and ineffective' state water enterprises, to introduce innovations and to install effective regulation that would improve the performance of the sector as a whole. In whatever form the reform may be implemented, the intended objectives of the reform have been closely associated with several key concepts which include (but are not limited to) efficiency, effectiveness, equity, competition and unbundling. Understanding these concepts is important not only to assess why reform was (or is yet to be) carried out, but also useful for explaining how they influenced state policy making. The following section presents an overview of the literature on water supply reform.

### 2.2.2 Understanding water supply reform

Different countries have adopted different approaches to water supply reform. One of the most popular approaches has been to promote private-sector participation (Kessides, 2004). This trend began to accelerate in the 1990s, and various forms of privatization (including service contract, management contract, lease, concession, BOT, BOOT, and divestiture) have been introduced in developing economies (Asian Development Bank, 2000; Kessides, 2004). However, some countries have embraced privatization under different paths such as Public-Private Partnership (also known as the Private-Financial Initiative in the UK) (Harris, 2003; Prasad, 2006; Fuest & Haffner, 2007), or Public-Private-Community Partnership (Franceys & Weitz, 2003). Another alternative is New Public Management (NPM). In contrast to privatization, NPM keeps the management and ownership of the water utility within the public domain, but introduces private sector management practices (Schwartz, 2008).

Even though the approaches to, and forms of, reform differ, the fundamental objectives are usually similar: improving the water supply system through greater efficiency and effectiveness in every aspect of the system. Reform takes place from different starting points, which require different approaches or (policy) interventions by the state. Some states directly involve private parties in (all aspects of) water management from the beginning, while others see gradual implementation (i.e. NPM) as a more feasible option than a full-fledged concession. It is clear that there is no such thing as 'one size fits all' approach. Careful selection of various approaches and a continuous examination of critical success factors can increase the success of a reform in meeting the desired objectives. As such, it is reasonable to believe that each country will embrace a form of reform most suited to their objectives and situation. For instance, some developing countries have embraced privatization to address financial constraints and efficiency issues faced by state-owned water enterprises (Cook & Uchida, 2008; Ehrhardt & Janson, 2010). Fuest and Haffner (2007) argue that privatization was often chosen to address the inefficiency and ineffectiveness of public water authorities and their inability to deliver an appropriate level of service or respond to the growing demands posed by rapid urbanization. The private sector is seen as more capable of tackling water-related issues such as non-revenue water, accessibility, quality and customer services. However these claims are disputed. Several scholars have highlighted cases where privatization has failed to improve efficiency or effectiveness, reduce costs or tariffs or expand coverage (Holland, 2005; Araral, 2009). They cite examples where private operators are only interested in serving rural areas where both affordability of water and willingness-to-pay among water users are relatively high. As a result few rural areas benefit from receiving a viable water supply. Political interference can also hinder network expansion. Private operators need revenues derived from tariffs to expand their networks especially to less profitable areas.

However, increases in water tariffs are politically unpopular and as such often opposed or resisted by politicians. On the other hand, politicians may see political mileage in promoting network expansion but also often refuse to grant or support the tariffs increases required to fund such expansion.

Some countries, having seen the failures of market liberalization and privatization, have opted to strengthen their public water authorities by introducing a reform to allow the institutional arrangements and 'management principles and practices associated with the private sector' (to be) introduced in the institutional context in which public water utilities operate (Schwartz, 2008: 49). This approach is referred to as New Public Management (Schwartz, 2008) and usually involves public water utilities being managed as a 'private sector' company, either through commercialization or corporatization (Rouse, 2007; Fuest & Haffner, 2007).

Another key goal of many water supply reforms is to promote good governance. In the water sector this can only be achieved when there is a separation between policy, regulation and service delivery (Ancarani & Capaldo, 2001; Rouse, 2007). In effect this usually entails the state having responsibility to formulate policy, regulation being carried out by an independent regulatory body, and service delivery by a water utility. In most circumstances, the regulatory body is created by the state to oversee the behaviour and performance of water utilities. Prasad (2007a) argues that regulation is necessary to address problems arising from lack of competition in the water industry and also prevent market abuse by private water utilities, since piped water supply is a natural monopoly (Nigam & Rasheed, 1998; Silvestre, 2012).

### 2.2.3 Explanation of key concepts

*Low and high level equilibrium*

Spiller and Savedoff (1997) describe the urban water sector in many developing countries as having a 'low-level equilibrium', in which low operational efficiency leads to low quality service and low willingness to pay by consumers. Many international organizations and donors, such as the World Bank, view private-sector participation (PSP) as a means of breaking this 'low-level equilibrium' and moving up towards 'high-level equilibrium', a situation characterized by high efficiency and service quality, and a high willingness to pay tariffs that at least cover the costs. However, PSP mainly seems to have positive effects on efficiency with small firms operating in competitive and unregulated markets. In the case of a water sector, where big corporations operate as a monopoly in a heavily regulated environment, the efficiency gains that can be made are less clear-cut (Anwandter & Ozuna, 2002). Lobina and Hall (2006) argue that private contracts may not lead to an improvement in the management and operation of water supply. They cite water supply privatization in the city of Grenoble, France, where a private sector contract was cancelled and reverted to municipal control. This was due to persistent price rises for consumers, questions about the legality of operations, and years of political activity against privatization. Such an example, where PSP has failed to improve efficiency and effectiveness, provides an argument for examining the potential for reforming public water authorities (the state, local governments, municipalities) into efficient and sustainable water supply providers (Lobina & Hall, 2006). Prasad (2006) assessed the results of 15 years of privatization in the water sector and concludes that PSP has had mixed results, and there is no clear evidence that the private sector outperforms the public sector. He attributes this to the 'economically flawed' and 'politically difficult' implementation of PSP. Kikeri and Nellis (2004: 87) are also sceptical about whether privatization produces the 'macroeconomic and distributional gains equivalent to its microeconomic benefits'. This has led to situations where governments have often been forced to step-in and take over the supervision of public (water) enterprises themselves. Jerome (2006) attributes the slow pace of water supply privatization in South Africa to the absence of an institutional framework: the lack of clarity about roles within and between government departments and the state-owned enterprises involved in privatization.

*Efficiency, effectiveness and equity*

Many developing countries embark on water supply reforms, such as PSP, to achieve three policy objectives: efficiency, effectiveness and equity. This is based on the argument that state-run water utilities only provide a third rate service (Nickson, 2002). They are seen as powerless when it comes to tariff setting, and are usually financially supported by central government. Many governments, especially in transition economies, regard water tariffs as a highly political issue. They are heavily subsidized, and fall far short of the true cost of supplying water. Heavily dependent on a (limited) budget from central government, and (low) tariffs, which cannot sustain their operations; most public water authorities have difficulties in maintaining their existing infrastructure, let alone undertaking service expansion and improvement. As a result, many governments have turned to the private sector for assistance.

It is generally accepted that PSP can lead to operational and financial efficiency by introducing new ways of managing business. This includes implementing tariff reforms that are based on volumetric and cost-reflective charging, cost reduction through pragmatic personnel policies, and establishing a pro-active customer complaints mechanism. In this way the private sector can provide services at the lowest possible costs, without compromising the quality of services (Kessides, 2004). In turn, consumers will benefit from lower tariffs in the long run.

Customer service and efficiency gains are also said to be the main forces behind the emergence of multi-utilities that offer more than one utility service (Kuks, 2006). In the Netherlands for example, a multi-utility company like Delta N.V. combines the supply of drinking water, with electricity supply, waste collection and cable services (Klostermann & Cramer, 2007). At the global level, several multinational utility companies such as Vivendi, Suez, Veolia and Lyonnaise des Eaux (France) and Thames Water (UK) have long ventured into water utilities management in developing countries, mainly through forming alliances with local partners (Kuks, 2006; Prasad, 2006).

The private sector is seen as having the potential to improve the effectiveness of water supply in terms of the availability of water, water quality and consumer satisfaction in general. As profit-maximizing business entities, they will strive to reduce monetary losses incurred from leakages, poorly maintained networks, and poor handling of consumer complaints. This will result in a wider coverage of people, improved water pressure and water quality due to fewer leakages and responsiveness to customer complaints (Nickson, 2002). However, without an adequate and well-defined regulatory framework, it is questionable if the intended objectives can be achieved by private services, and if so, at what cost for consumers (Nigam & Rasheed, 1998; Pongsiri, 2002). Effective regulation is a prerequisite, not only for ensuring the sustainability of the sector in the long run, but also to curb market abuse by privatized operators in a situation where there is an absence of competition (Nigam & Rasheed, 1998). Yet as Fuest and Haffner (2007) argue (political) interference can also, undermine the ability of the regulator to perform its tasks independently.

Another main objective of water supply reform is to provide equal access to water to all, especially to the poor who are often not connected to the public water system. To achieve this, the structure of water tariffs has to take into consideration the ability (of poor consumers) to pay for services, and establish a progressive tariff regime to discourage wasteful consumption. Tariff reform is crucial to promoting the efficient use of water (Asian Development Bank, 2000). However, customers are more willing to accept gradual rather than sudden tariff increases. Full-cost recovery pricing is required to generate the revenue needed to allow water operators to expand existing networks to areas that are not yet served. Yet, high connection charges and cost-recovery tariffs might prevent the poor from seeing any benefit from network expansion. To overcome these problems creative financial arrangements may have to be explored. These can include a direct subsidy scheme by the government, or staggered payments of connection charges by water operators to help promote access to water supplies.

*Competition in the market*

As a natural monopoly industry, the water supply market rarely involves direct competition (Kessides, 2004). It is unknown to have two piped water operators providing services in the same area. It is also hard to imagine water pipes of two different water companies installed in one neighbourhood or house. However, in large metropolitan areas, such as Paris and Manila, the water market is split into different service areas which have different operators although they do not directly compete with each other (Kessides, 2004). In the state of Selangor and Johor, Malaysia, water treatment operations have been commissioned to three different operators who supply treated water to a single water distributor (MEWC, 2004). This, however, is unusual and in Malaysia (and many other developing economies) the lack of competition in the sector is exacerbated by three factors. First, very long contract periods result in a *de facto* long term monopoly (Linjun, 2007) and the exclusive rights that these contracts provide make it difficult to bring in other competitors to the same service area (Kessides, 2004). Second, it is difficult for new entrants to enter the market. It takes a relatively long time for new entrants to undertake ancillary works including land acquisition, develop costly new assets and networks (distribution pipelines, dams, treatment plants, reservoirs, etc.) and to build-up the market for their services (Asian Development Bank, 2000). Third, according to Kirkpatrick, Parker and Zhang (2006), competition in the water market is usually cost-inefficient. The costs of pumping water into distribution networks are (proportionately) much higher than the costs of distributing energy, or transmitting telephone calls.

Competition in the water market can occur between piped and un-piped sources (such as vendors and wells), and between large water operators and small-scale operators. Consumers will turn to un-piped water, even at a higher price, when piped water is of a poor quality (Kessides, 2004). In Paraguay and many African cities, consumers rely on small-scale operators for water supply in areas that are not reached by large water operators due to an extremely low population density, difficult topography or chaotic layout. A study by Nyarko (2007) indicates that 40% of the urban population in Ghana relies on alternative service providers and often pay water charges between 5 to 14 times higher than the lifeline tariffs per cubic meter of the government water company.

Competition in the water market (and other network-bound utilities) can also happen when governments allow third party access to the sector. Third party access can be used as a platform to break up the monopoly of existing service providers. Competition can be further enhanced by breaking up (service) areas geographically and allowing multiple service providers. For example, a government can facilitate third party access by granting a new entrant access to an incumbent supplier's water infrastructure and services (Massarutto, 2007; Marsden Jacob, 2005). The third party access is believed to have a relatively high degree of success in promoting competition in two circumstances. First, where there is 'a natural monopoly market where it is more economic for a single facility or network system to supply the market than to duplicate the facility or network'; and second, where it allows 'a third party (access seeker) already has access to services provided by an access provider, the access seeker may apply to increase the scope of its access beyond that provided under the earlier access terms and conditions' (Marsden Jacob, 2005: ii).

However, competition through third party access can be severely jeopardised if the same services are contested by substitutes (Tasman Asia Pacific, 1997).

*Competition for the market*

As many developing and transitional economies open up their water supply sector to PSP, competition among private water operators to secure water contracts from the government is becoming more intense. Besides running long concessions, private water operators may also be involved in management contracts, service contracts, lease contracts, BOT and BOOT arrangements, joint ventures and divestitures (Nickson, 2002; Kessides, 2004). Management contracts are usually given out to private water operators for shorter periods (normally up to 5 years) and are limited to operations and maintenance. On the other hand, it is common for water operators to award service contracts on supply and civil works, technical assistance or sub-contracting or contracting out aspects of the water supply services such as meter reading, IT services, tariff collection, design, and even grass cutting (Asian Development Bank, 2000). Under a lease contract, a (public) water utility leases the full operation and maintenance of its facilities within an agreed geographical area to a private operator for a period of time. In exchange for a lease rental, a private water operator is granted the right to invoice and collect charges from customers within that area (Asian Development Bank, 2000).

Another form of PSP is a concession where a private operator takes over the management and operations of water supply provision from government (Asian Development Bank, 2000). A concession relieves the government from operations and raising capital, but more importantly, it is expected to increase the operating efficiency. A private operator managing the whole water supply system (both treatment and distribution), is exposed to certain risks: operational risks, financial risks, and collection risks (Othman, 2007). However, (full) concessions have been criticized for being politically impractical. For instance, there are cases where contracts have been awarded through political relations rather than on merit. In this circumstance, the presence of the political influence has undermined effective regulation (Asian Development Bank, 2000; Prasad, 2006). As such, BOT and BOOT are seen as alternatives to concessions, especially when much investment is needed for bulk supply operations (Asian Development Bank, 2000). Under BOT and BOOT, water operators are contractually obliged to finance the development of water infrastructure, manage it and return it to the government at the end of the contract period. In return for the investments made, the water operators are given the legal right to collect, retain and use the revenues generated from the provision of water. Normally under both types of contracts, a regulatory authority is established to regulate the conduct of water operators.

Water supply can also be developed through a joint venture arrangement between public and private water utilities, or by means of divestiture. Divestiture entails the direct sale of infrastructure to the private sector by selling assets or shares or through a management buy-out (Asian Development Bank, 2000; Kessides, 2004). It is arguable that concessions and BOT and BOOT constructions do not contribute much to raising competition in the water market, especially when only small and active competitors or bidders are involved (Kirkpatrick *et al.*, 2006). Around the globe few active multinational water companies (MNCs) such as Suez, Vivendi, Veolia and

Thames Water (Prasad, 2006) have become involved in water sector privatization, and become a dominant force in the entire water sector especially in developing countries.

## Unbundling

Ancarani and Capaldo (2001) propose to inject the working culture of the private sector into public water utilities to increase the efficiency level. This can include requirements for public authorities to award contracts through competitive tendering, or identifying services or activities that can be assigned to the private sector. One possible way of doing this is by unbundling. The primary objective of unbundling a utility sector is to promote competition. Unbundled management is likely to have a better focus on the capacity and productivity of the individual components and their interface with each other. Evidence from many countries also shows that unbundled infrastructures, where individual components are managed separately, perform better than centrally managed networks (Asian Development Bank, 2000). Unbundling water supply sector entails separating or breaking up the entire (water) network into several segments: (raw water) abstraction, treatment/purification, distribution/transportation, retailing, billing and revenue collection. Usually these unbundled activities are assigned to different parties to manage. Competition can then be introduced into these newly created business units, through measures that include effective competitive bidding and open tendering. Once the network has been split up, appropriate regulations must be enforced to avoid re-integration and to ensure that the industry retains a number of unbundled firms.

Unbundling in the context of BOT/BOOT operation has been considered less favourable than more vertically integrated structures (Rouse, 2007). It is less favourable because BOT/BOOT does separate the main activities – treatment, distribution and collection – which limit the ability of the management to optimize the (shared) resources available within different organizations. By contrast, vertical integration of the whole water supply chain has the strength of taking into account the relationship between costs, revenue collection and customer services. It allows water operators to share or mobilize the same resources in managing the entire water supply system: from raw water abstraction, treatment, distribution, billing and revenue collections, customer services, and even wastewater. Vertical integration favours the entire water supply chain being managed by the same organization, which can optimize its use of resources, and be aware of the cost structures and their relations. This model has also proved useful in facilitating 'cross subsidies' between different activities in the water supply system.

Even in cases where the utilities remain in public hands after unbundling, a pro-competitive reform can still be introduced through commercialization or corporatization. Unbundling also facilitates indirect competition through comparative competition (Asian Development Bank, 2000) or benchmarking (VEWIN, 2010). Comparative competition, which is widely used in the UK's water sector, allows the regulator to compare the performance of different water operators in aspects such as water tariffs by customer class, non-revenue water, the cost of producing one cubic metre of water, and response time to customer complaints. VEWIN (the Association of Dutch Water Companies) uses benchmarking to measure the performance of Dutch drinking water suppliers in four core areas: drinking water quality, service, environment and finance, and efficiency. Unbundling also can be used to stimulate benchmarking cultures among water operators in various segments of their activities (Dorsch & Yasin, 1998). For instance, it is often easier to compare and measure the performance of water utilities when the water cycle is broken into smaller business units – treatment, distribution and collection – earlier termed as horizontal unbundling. An operator may perform well in one area, but not necessarily in others. This approach allows water utilities to adopt a culture of 'learning from others' among themselves – share how they did well in some areas, and learn why others excel in the other areas. This measure is designed to put pressure on under-performing water operators to improve their overall efficiency and effectiveness, or having to face rigorous public scrutiny. Usually the role of carrying out and enforcing benchmarking has been the key task of the central regulator. Nevertheless, it is argued that the task of the regulator can be severely hampered without the presence of reliable information. It is here where the regulator has to be equipped with certain powers to do three things: to acquire information from water utilities; to undertake information audits to determine authenticity and credibility; and to publish the performance of the companies to the wider public.

These concepts provide insights into the rationalities that underlie water supply reform. Understanding these key concepts also helps us to understand how the state can better manage its relationships with non-state actors, an important aspect of water supply governance. Water supply reform can take different forms and approaches. The state's choice about which form to adopt will probably depend on the given circumstances. This section has also shown that water reform requires the state to undertake different structural changes, and put in place a set of (water) laws that support (various forms of) private sector participation and new public management. One of the factors underpinning successful reform is the existence of a central regulator which is able to undertake regulation independently and free from (political) interference. In many developing countries, including Malaysia, there is no guarantee that this condition will be complied with and there is always the possibility that water regulators may be subject to political interference.

## 2.3 Water supply reform: theoretical perspectives

Water supply reform in developing countries can be viewed as a state policy intervention, designed to achieve certain policy objectives, such as efficiency, effectiveness and equity in water supply. This was the case in Malaysia where there was a determination to remedy and improve the existing water supply situation and to ensure the attainment of long-term national development goals. The formulation of public policy (in this case the reform) involves a series of processes. By analysing these processes we can get a better understanding of the outcomes of the reform and their impacts on target groups (Fischer, 2006). The policy arrangement approach offers a useful analytical tool to analyse and understand day-to-day policy processes (Liefferink, 2006; Arts & Goverde, 2006). In the quest to understand and analyse the policy process, several questions need to be asked. Who are main actors in the sector? What rules govern their relationships? Who controls the availability of resources? How do actors use their power to mobilize resources? And how do the dominant actors seek to shape the policy discourses? The policy arrangement approach, developed by Van Tatenhove, Arts and Leroy (2000), provides an analytical framework to answer these questions. The next section explains the concept and its applicability to the water supply sector.

### 2.3.1 The policy arrangement approach

The policy arrangement approach is derived from a policy sciences framework developed by Lasswell (Arts & Van Tatenhove, 2004). Van Tatenhove *et al.* (2000: 54) define policy arrangements as 'the temporary stabilization of the content and organization of a particular policy domain'. It refers to the way in which a policy domain is shaped, in terms of substance and organization, in a bounded time-space context (Van Tatenhove & Leroy, 2003; Arts & Van Tatenhove, 2004). Arts and Van Tatenhove (2004) argue that substance (principles, objectives, measures) and organization (departments, instruments, procedures, division of tasks and competence) are two important aspects of policy development. Policy domains are dynamic and develop in response to the constant pressure for change and policy innovation. Policy arrangements may also evolve at different levels of policy-making – local, national and transnational. Liefferink (2006) and Arts and Van Tatenhove (2004) see the policy arrangement approach as a tool for understanding and analysing day-to-day policy processes through four dimensions:
1. Actors and their coalitions involved in the policy domain.
2. The division of resources between these actors, leading to differences in power and influence (where power refers to the mobilization and deployment of available resources, which influence who is able to determine policy outcomes and how).
3. The rules of the games currently in operation, in terms of the formal procedures of decision making and implementation as well as informal rules and 'routines' of interaction.
4. The current policy discourses, where discourses entail the views and narratives of the actors involved (the norms, values, definitions of problems and approaches to the solution).

The first three represent an organizational dimension of a policy arrangement, whereas the fourth one refers to the substantial aspects of policy. A schematic overview of the policy arrangement approach is shown in Figure 2.1.

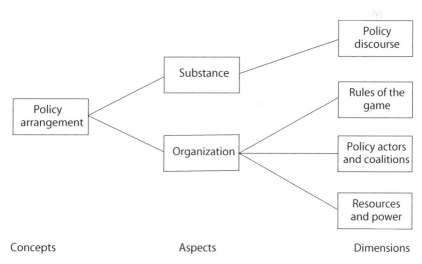

*Figure 2.1. Concepts, aspects and dimensions of policy arrangement.*

Liefferink (2006) argues that these four dimensions have been developed from a series of theories and approaches from sociology, public administration and political science (such as network theory, and discourse analysis). The four dimensions are interconnected, and understanding their interrelatedness can help us better understand a policy arrangement at any given point in time. To illustrate this interconnectedness, Liefferink (2006) represents them as a tetrahedron, with each corner representing one dimension (Figure 2.2).

The figure below implies that any attempt to analyse policy arrangements needs to address the entire tetrahedron, as they are 'inextricably interwoven' (Liefferink, 2006: 48). As such, changes in one of the dimensions are likely to induce changes in the others. This is because 'there is no such thing as a policy without substances (principles, objectives, measures) and equally there is no such thing as policy without organization (departments, instruments, procedures, division of tasks and competence)' (Arts & Van Tatenhove, 2004: 341). For instance, a decision by the state to open up and liberalize the water market will result in the entry of new actors/players into the market, or force the existing actors to form coalitions so that they are in a stronger position to mobilize existing resources (and acquire new ones) in order to remain in business. The existing rules (of the game) might have to be altered, and new rules may need to be written to ensure healthy competition among actors, and protect consumer interests. Newly emerging policy discourses in the water sector policy domain demand the establishment of new rules that recognize water as both an economic and a social good. However, this does not mean that all parties will accept the establishment of these new rules. Some might contest the rules, if they do not work to their advantage. This highlights that the interconnectedness of the four dimensions of policy arrangements does not necessarily imply that there is harmony, stability or internally consistence between them. For instance, existing actors or players in the sector might not totally agree with new rules set by the state to liberalize the water market. New actors might be seen as 'enemies' eroding the long standing presence of existing operators in the market and jeopardizing their business. Established players in the policy domain might resort to using political connections to counter the new 'rules', to undermine, or even get rid of the new entrants (Liefferink, 2006). Equally if civil society (particularly consumer groups and environmental movements) feels threatened by policy changes they may join forces to counter this threat. This may force the state to re-examine

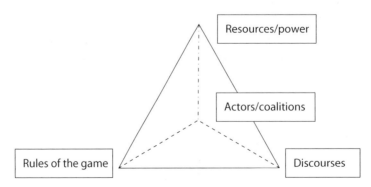

*Figure 2.2. The tetrahedron, symbolizing the interconnectedness of the four dimensions of a policy arrangement.*

the proposals and policy discourse, and even consider institutional change in order to 'renew the day-to-day interactions in arrangements' between the state, market and civil society (Arts & Van Tatenhove, 2004: 344).

### 2.3.2 *Applying the policy arrangement approach to the water supply sector*

Water supply reform implies a certain degree of institutional change, (new legislation, procedures, institutions; and power relations between actors in the domain) and the policy arrangement approach provides a useful framework to observe these changes. The reform alters the old rules and introduces new ones, which will guide the long term direction of the sector and govern the behaviour of the involved actors. It gives rise to new discourses that will affect the way actors mobilize, share, control, or re-distribute resources (in order maintain their power). Such changes not only involve institutional reform, but also the need for the state to re-invent its traditional role because of the growing importance of the role of non-state actors (Spaargaren & Mol, 1992; Mol & Spaargaren, 1993; Mol, 1995). This is one of several factors likely to provoke 'deep' institutional change (Wiering & Arts, 2006: 337), where policy, regulation and service delivery are separated and placed under different, existing and newly-created, institutions. The analysis in this volume will focus on the four dimensions of policy arrangement approach – actors and coalitions, resources and power, rules of the game and discourses – to analyse the policy-making process of water supply reform in Malaysia. It will seek to identify the differences in policy arrangements for water provisioning that existed before and after the reform. These mostly affected the division of power among stakeholders, the rules and routines of the game and a change in the dominant discourse. Below these four dimensions will be discussed first from a broader theoretical perspective, citing examples from the policy domain of water sector (reform) in different countries.

### *Actors and coalitions*

This first dimension of policy arrangements focuses on actors and their coalitions in the policy domain. Liefferink (2006) argues that most policy studies begin with identifying the actors involved in the relevant policy domain. The discussion on the other three dimensions – rules, resources and discourses – will only materialize after the question about who is involved has been answered. This step involves identifying the relevant actors and their influence in the policy process. Broadly, we can distinguish between state and non-state actors. The state actors represent government agencies – ministries, federal and state departments, regulators, etc. They usually have a close interface with non-state actors such as market/economic actors, firms, non-governmental organizations (NGOs) and other non-profit organizations. Liefferink (2006) regards the interaction or relationship between state and non-state actors as a critical characteristic of governance. In the course of a policy arrangement, new actors often enter the domain, or existing actors retire from it. Actors might act individually or in coalitions in trying to influence the 'rules' of the game that guide the policy process. They tend to act individually if they have the power to mobilize resources – money, technology, expertise, etc. – by themselves. If no single party dominates or controls the available resources, it is more likely that shared power coalitions will need to be established. In this situation, actors will work in coalition to share, control, distribute, and re-distribute resources.

However, it is not guaranteed that a coalition among actors will remain in place. Coalitions may fall apart for numerous reasons, e.g. when a clash of ideology or ideas cannot be avoided, or when a coalition becomes irrelevant to the policy discourse. A coalition may also be dissolved when it has accomplished its mission. Equally it may stay together, channelling its synergy to address other policy problems or discourses, or achieve other objectives.

We can now turn our attention to the actors and coalitions in the water supply sector. It has been recognized that water supply is predominantly a public domain. Around the world, water supply has usually been managed by government departments, municipalities, water boards, or state-owned enterprises. State actors are traditionally the main players in the water supply policy domain (Rouse, 2007). The government decides which rules to apply and enforce, who has the power to mobilize resources, and what norms, values and solutions are used in the policy discourse. In a federal system of administration (as in Malaysia), the policy discourse is clearly dominated by the Federal government. It is in a stronger position and has the power to mobilize recourses, set the rules, etc. Coalitions among state governments are unlikely to have major effects on the outcome of policy arrangements made at the federal level, as state governments are dependent on federal assistance for water projects. However, when the private sector involvement started to get involved in the water sector in late 1990s, it opened up an avenue for economic actors taking over responsibility from state actors in water supply provision. This forced the state to write new 'rules' not only to accommodate the new policy arrangements, but also to safeguard the right to water for its citizens. The potentially lucrative returns from the water sector attracted many new actors to the sector, (who are viewed by existing actors as a threat to their existence). Faced with this situation the state needs to introduce 'new' rules in the form of legislation, regulations and procedures, and political culture (political rules of the game) to curb any possible market abuse by market actors (Wiering & Crabbe, 2006). Markets actors are confronted with questions of how they can best mobilize and make use of their resources to meeting the 'new' rules (of the game) imposed by the state.

*Resources and power*

The second dimension of the policy arrangement approach is about resources and how they are allocated. Actors and their coalitions need to have resources – money, technology, expertise, etc. – to influence and intervene in policy decisions that affect certain policy domains. The ability of actors to mobilize resources largely reflects their dominance (or otherwise) within a policy arrangement. We can argue that whoever controls the resources is in a better position to control the outcomes of policy decisions. Logically all actors would like to be in the position, where they can decide the rules of the game, who should be allowed to enter the policy domain, be in (or out of) coalitions, have access to resources, and influence the values and norms of the policy discourse. However, in general no single actor entirely dominates or controls the distribution of all resources. Resources – money, personnel, knowledge, technology, expertise – are differently distributed among actors. In a given policy arrangement such as water privatization, state actors might be well equipped in terms of legal resources and water rights, while market actors may be more capable in securing financial assistance from the capital market, and assembling expertise through the power relations they have with other actors. They also depend on other actors for resources. Unequal distribution and non-exclusive control over resources, makes it inevitable that some actors will be driven to forge a coalition with other actors (resource coalitions) (Liefferink, 2006). Such coalitions lead actors to share resources, and to a certain degree, initiate a re-distribution of resources to correct the imbalance of distribution.

Resources can be used both as 'weapons' and as 'prizes' by actors (Rhodes, 1986). As weapons, actors use resources to determine the outcomes of policy decisions. On contrary, resources can be used as prizes to lure others to his side and thus allowing it to influence and change the distribution of resources to its advantage. Actors need power to mobilize resources in order to act and intervene. In this circumstance Giddens (1984) sees the concept of power as the capacity of agents to achieve outcomes in social practices. Arts and Van Tatenhove (2004) argue that the capability of actors to mobile resources is restricted by an unequal distribution of resources among actors in the policy arrangement, due to differences in power. This line of argument emphasizes that the concept of power is closely linked to the concept of resources.

Actors rely on power to mobilize resources and their ability to share, control and distribute resources is also dependent on their power relations with others. Arts and Van Tatenhove (2004: 347) define power as 'the organizational and discursive capacity of agencies, either in competition with one another or jointly, to achieve outcomes in social practices, a capacity which is however, co-determined by the structural power of those social institutions in which these agencies are embedded'. According to them, actors have the potential to influence the development of policies in a policy arrangement, as well as the impact on the structural context in which they operate. More specifically, it provides insights into how change (in a policy arrangement) is linked to power. Arts and Van Tatenhove (2004) have developed a three-layered model to analyse the linkages between change and power: relational power; dispositional power; and structural power (Table 2.1).

Relational power refers to the situation where an actor uses power to achieve policy outcomes through interactions with other actors. Usually actors can use relational powers to achieve outcomes in two ways: first, where it is used against the will of actors (transitive), and second, where it serves as a mean to working together with actors (intransitive). Dispositional power refers to the positioning of actors in an organization, through which they attempt to achieve their goals through the use of relational power (Arts & Van Tatenhove, 2004). In this layer, power is mediated by rules and resources: rules define and legitimate the position of actors within the organization, whereas resources determine the autonomy and dependence of individual actors and their position

Table 2.1. Three layers model of power (Arts & Van Tatenhove, 2004).

| Type of power | Focus |
| --- | --- |
| Relational (transitive and intransitive) | Achievement of policy outcomes by agents through interactions |
| Dispositional | Positioning of agents in arrangements mediated by rules and resources |
| Structural | Structuring of arrangements mediated by orders of signification, domination and legitimization |

within the organization. Structural power can be defined as 'the way macro-societal structures shape the nature and conduct of agents, being both individuals and organizations' (Arts & Van Tatenhove, 2004: 350). Structural power works through discourses and organizations (political, legal, economic), which are used by actors to determine the legitimacy (or otherwise) of acts or thoughts. Agents operate with 'structured asymmetries of resources' and this determines their ability to achieve desired outcomes in their social relationships. Arts and Van Tatenhove (2004) argue that structure and organizations cannot determine the conduct of agents.

What resources and power relations might one expect to find in the water supply sector? In a traditional state monopoly environment, all the resources and power (executive and legislative) with regard to water supply provisioning are vested in the state. This gives the state almost 'resource exclusivity', giving it the power to set the rules of the game, to decide about the values and norms of discourses and which actors can enter or withdraw from the domain. Only rarely in negotiations with other actors, is the dominant role of the state questioned (Pierre & Peters, 2000). However, the introduction of market liberalization partly compromises the state's dominance and its resource exclusivity. The state enters into this new position voluntarily in order to achieve certain policy outcomes. The state is become dependent on other societal actors because it lacks sufficient resources to deliver a public service. This introduces a new dimension to the policy arrangement in which transitive power might not always work, but where working in collaboration with other actors, (intransitive power) seems to offer prospects of achieving desired outcomes.

The implementation of PSP water projects demonstrates the relation between powers, actors and distribution of resources. For instance, state governments forge collaborations with private actors which involve assigning the risks to the party who can best manage them (intransitive power). Usually private actors are seen as more capable in managing financial risk. On the other hand, BOT/BOOT constructions are seen as giving the state more capacity to mobilize 'legal resources' – legislations, regulations, orders, etc. – by working closely with other state actors. The same scenario puts the private sector in a better position to mobilize or acquire other resources – money, expertise, knowledge. The collaboration between the two will create the synergy needed to achieve desired policy outcomes. This situation of resource interdependency leads to new power relations emerging between the actors involved. However, the conduct of actors, especially of market actors, must be based on a set of rules to curb the possibility of market abuse in the absence of competition. Normally, the power to dictate rules is retained by the state, and institutionalized in the form of a regulatory authority. In the event where regulatory power is weak, there is a tendency for market actors, based on their position, and power relations, to seek outcomes at the expense of other actors (transitive power). They may do this, for instance, by influencing the state to formulate rules that work to their advantage, or withdrawing or introducing certain resources (a tax, levy or incentive) that might make it unattractive for new actors to enter the market. On the other hand, access to new knowledge (resources) can help civil society to question if the state is protecting the interests of consumers and the environment. Consumer and environmental groups may form a coalition to lobby the state to recognize the importance of customers' concerns and ecology when changing the rules that govern water supply (Pierre & Peters, 2000).

*Rules of the game*

Liefferink (2006: 56) defines rules as 'mutually agreed formal procedures and informal routines of interaction within institutions'. According to Buizer (2008), there are two categories of rules: formal and informal. Formal rules are rooted in legal texts and documents. Boot (2007: 48) defines formal rules as rules that are 'written down and approved of by the majority of stakeholders or, at least, the most powerful ones'. Informal rules normally represent the 'dos' and 'don'ts' of a political culture. Usually these 'rules are recognized by society, but they are not officially recorded as rules such as norms, behaviours or agreements' (Boot, 2007: 49).

Rules operate in two ways. They *enable* actors and coalitions to achieve outcomes in social interactions. Yet they also *constrain* the conduct of actors. As clearly shown in the tetrahedron (Figure 2.2), a change in rules will trigger changes in the other dimensions, with which they are closely interconnected. In a policy arrangement, rules mediate actors' interactions. When rules merge with resources and power, regulatory power is created (Liefferink, 2006). We might want to identify what discourses underlie the rules of interaction that prevail in the network, especially those that determine interactions between the state, market and civil society.

Liefferink (2006) suggests that the rules dimension of policy arrangements can be used to study the influence of institutional change on particular policy areas, and the rules dimension of policy arrangements fits well in analysing institutional change in the water sector. In the early 1990s, there were drastic institutional changes when many countries started to liberalize their water sector. Following the introduction of a privatization policy, the state introduced new 'rules' to govern the interactions between actors within the water sector policy domain. Legal texts and documents were written and presented to Parliament for consent. Formal procedures, which facilitate the separation between policy, regulation, and service delivery, were drafted. Independent regulatory bodies were institutionalized. To a certain degree, these institutional changes affected the interactions of actors and coalitions in the given policy arrangement. For instance, water operators will have to alter their behaviour to comply with new 'rules' introduced by the regulator. Equally, in discharging its duties a regulator is bound to consider the existing rules of the government, the spread of privatization, and rules or norms imposed by international donor organizations.

It is also important to examine how the state, market and civil society actors negotiate with each other in meeting their needs. Normally, the relationships between these actors are guided by informal rules. Such rules include (informal) agreements and procedures, and they are just sometimes based on mutual interests and trust (Boot, 2007). Sometimes they exist in the form of internally enforced standards of conduct such as ideas, ideologies and choice. Whatever form they may take, informal rules dictate which parties are allowed to take part in the policy process, which actors can act as nominees to those actors not allowed at the negotiating table, the 'do´s and don't´s' of the meeting, and what discourses take centre stage during the policy debate. In most cases when the supporting discourse fades away, these rules automatically disintegrate. When new discourses emerge, a new set of informal rules are established. An amalgamation of (several) actors might take place in response to these rules. For instance, the civil protest against the Chochabamba water concession in Bolivia was led by the most vulnerable section of society who would have been the worst affected by this project (Casarin *et al.*, 2007; Schwartz, 2008). They were brought together by their perceptions that the private water sector was only interested in serving the rich, that they would be deprived of their lands, and that they would have to pay for water (Holland, 2005).

*Discourses*

The last dimension of the tetrahedron is discourses. An early definition of discourses can be traced back to the work of Hajer (1995: 44) who defined them as 'a specific ensemble of ideas, concepts, and categorizations that are produced, re-produced and transformed in a particular set of practices and through which meaning is given to physical and social realities'. Building on this work, Arts and Van Tatenhove (2004: 343) re-define policy discourses as 'dominant interpretative schemes, ranging from policy concepts to popular story lines, by which meaning is given to a policy domain'. Liefferink (2006) argues that policy discourses are relevant at two different levels. At the first level, they consist of general ideas about the relationships between the state, market and civil society, which may influence a specific policy arrangement. The second level concerns ideas about the concrete policy problems at stake – their characteristics, the nature of problems, their causes, and possible solutions. This dimension of discourses is intrinsically linked to the concept of power, especially when we recall that an actor or coalition needs power to mobilize resources (Arts & Van Tatenhove, 2004). Through power, resources can be mobilized and directed to solve concrete problems and to prevent re-occurrences.

In the water sector, there are on-going policy discourse debates between the need to retain public water authorities, or to open up water markets to private sector involvement. The arguments levelled against public water authorities point to their incapability to deliver water effectively and economically and the need, therefore to shift these responsibilities to the private sector. Others are of the view that water must not be treated as commodity, and should remain in public hands. As such, the discourse story line revolves around questions like: 'what causes the public sector to be inefficient?' and 'what (possible) solutions are available to turn around public water authorities?' One possible solution is to transform public water authorities into adopting the principles of New Public Management by introducing the institutional arrangements and working cultures associated with private sectors into the public domain (Schwartz, 2008; Polidado & Hulme, 1999). But when privatization is chosen as the preferred option this introduces a different perspective. The issues at hand now are how the state can mobilize resources to facilitate competition among private water operators, which might eventually make the sector sustainable in the long run. The ability of the state to mobilize resources depends on the 'regulatory power' it derives from legal texts and documents – legislation, regulations, statutes, etc. An example is the legislative 'overhaul' which marked a change in the way the Dutch government dealt with water and flood defences: reflecting a discursive shift from 'the battle against water' to new discourses of 'accommodating water' (Wiering & Arts, 2006: 328).

I have explained in detail the four dimensions of policy arrangements, and their application in relation to the water supply sector. The policy arrangement approach can be a suitable model for explaining and analysing the policy process of water supply reform. Yet the policy arrangement approach does not provide the tools to assess whether the policy actually leads to attainment of the desired goals, or what the impact of reform is on the behaviour of water operators. As the policy arrangement approach does not examine the goals achieved by the water reform process, another approach, a policy evaluation approach, is required to evaluate whether state interventions attain their set goals. A policy evaluation analysis is a 'scientific analysis of a certain policy area, the policies of which are assessed for certain criteria, and on the basis of which recommendations are formulated' (Crabbe & Leroy, 2008: 1). This definition contains several fundamental elements of policy evaluation: the concept of policy, (policy) analysis and evaluation, the criteria for policy analysis and evaluation, and recommendations to improve policy. As such, a policy evaluation approach should analyse and evaluate the impacts of policy interventions. The following section (2.4) explains policy evaluation approaches in more detail, and its relevance to evaluating water supply reforms.

## 2.4 Policy evaluation approaches

### 2.4.1 Introduction

I have discussed at length the four dimensions of policy arrangements, which will be used to explain and analyse the policy processes that occurred in Malaysia's water supply reform. Such a policy arrangement approach helps to identity precisely who (which actors) is (are) involved in the sector, how they mobilize resources, what rules are being observed, how this affects the entire policy arrangement, the power relations between actors, and how these affect the policy discourse. Evaluating the impacts of water supply reform on the performance of the sector needs a different framework for analysis. To evaluate the performance of water operators in two areas (operational efficiency and environmental effectiveness) for a period *before*[4] and *after*[5] the reform a policy evaluation approach is developed from the European Environment Agency's (EEA) policy evaluation model, the American Water Works Association's (AWWA) QualServe Business model and the Malaysian Water Association's (MWA) model. The EEA framework presents a policy evaluation framework to assess the impacts of water supply reform, while the AWWA and MWA models are used to provide indicators for measuring operational efficiency and environmental effectiveness.

Before presenting and discussing the policy evaluation approach in detail, it is important to discuss the concept of (policy) outcomes, what policy evaluation can contribute to policy analysis, and difference between *ex ante* and *ex post* policy evaluation.

### 2.4.2 Policy outcomes

According to Fischer (2006), the evaluation of public policy can focus on policy or programme outcomes (or impacts), or on the processes by which a policy or programme is formulated and implemented. Dunn (2004) suggests that we can observe two kinds of policy outcomes: outputs, and impacts. (Policy) outputs are 'the goods, services, or resources' received by target groups or beneficiaries, while (policy) impacts are 'actual changes in behaviour or attitude that result from policy outputs' (Dunn, 2004: 280). Dunn also distinguishes between target groups and beneficiaries. Target groups consist of individuals, communities, or organizations whom a policy or programme

---

[4] 2004-2006.
[5] 2007-2009.

is designed to influence, while beneficiaries are groups which benefit from a policy or programme. Crabbe and Leroy (2008: 5) view policy outputs somewhat differently, as both the 'quantity and quality of products and services delivered by policy makers'. They see a policy outcome or social change in terms of the behavioural change among target groups (individuals or organizations) that occurs as a result of policy implementation (Crabbe & Leroy, 2008). From these two definitions, we define policy outputs as the tangible outcomes experienced by the target groups as a result of a policy intervention. Policy outcomes, on the other hand, concern how the policy outputs of a policy intervention induce behavioural change within the target groups. This is very much in line with how the EEA model defines outputs and outcomes.

### 2.4.3 *The purpose of policy evaluation*

An evaluation of public policy can serve three purposes (Dunn, 2004). The first, and may be foremost, is to assess *policy performance*. This involves examining whether the particular goals or objectives of a policy have been attained. Second, in the event that a policy failed to achieve its desired objectives, policy evaluation allows policy makers (and others) to *clarify and critique* the selected goals or objectives and their underlying justifications. This process can lead to a questioning of the appropriateness of the goals and objectives. It can also provide a platform for examining alternative goals and objectives (Vaz, Martin, Wilkinson & Newcombe, 2001). Third, a policy evaluation process can contribute to reforms and recommendations of policies. As time goes by, existing policy problems need to be re-structured, which involves re-defining policy goals and objectives. In circumstances where existing policy has repeatedly failed to produce the desired or intended results, even after rounds of reviews, policy makers can use evaluations to recommend new policies or reform existing ones. This might lead to the abandonment of a previously favoured-policy, its amendment or its replacement by another one.

### 2.4.4 Ex ante *and* ex post *evaluation*

*Ex ante* evaluation is 'a forward-looking assessment of the likely future effects of new policies or proposals' (Vaz *et al.*, 2001: 9). In other words, a policy is evaluated before being implemented (Crabbe & Leroy, 2008). This can take place at different levels of activity (e.g. policy, programme, or project) (Vaz *et al.*, 2001). For example, *ex ante* evaluation can be used to predict whether implementing particular water and sanitation policies will help developing and under-developed countries to meet the United Nations' Millennium Development Target 10 (reducing by half the number of people without sustainable access to safe drinking water and basic sanitation by 2015). *Ex ante* evaluation can also be used to assess the cost-effectiveness and benefits of alternative programmes before choosing which one to implement. In contrast, *ex post* evaluation focuses on 'what has actually happened following the introduction of a particular measure' (Vaz *et al.*, 2001: 9), or after a policy has been implemented or developed (Crabbe & Leroy, 2008). In general terms, it refers to the assessment of the actual effects of a given policy, programme, or project. Policy makers can use the information gathered from *ex post* evaluation to gauge the impacts of a given policy or measure, and learn from both the successes and failures of that policy or measure (policy learning). For instance, policy makers (from developing economies) could engage *ex post* evaluation to measure the effects of water privatization in general. This can provide a valuable learning experience for policy makers, which might prove useful for designing (*ex ante*) a more comprehensive policy programme in the future.

Given the nature of this research, which analyses the impacts of the on-going reform of the water supply sector, neither *ex ante* nor *ex post* evaluation seemed to be wholly appropriate. Instead, this policy evaluation falls somewhere the two. It is, in the words of Crabbe and Leroy (2008), an *ex nunc* evaluation. The effectiveness of an on-going public policy process can be measured at least at two intervals: current, and interim (which may lead to policy modifications). In this sense it could be argued that there is no real *ex post* evaluation of public policy (as all policies are on-going unless completely abandoned). *Ex post* evaluation can, however, help describe what actually happened following the implementation of a policy and also what we can learn from this policy intervention. Policy makers can benefit from this 'policy learning' to assist them in formulating future policy interventions, or potential alternatives (*ex ante*). Policy makers can even modify or alter the implementation of policy to ensure the attainment of desired policy outcomes.

### 2.4.5 The European Environment Agency's policy evaluation model

The European Environment Agency (EEA) has developed a policy evaluation framework to analyze the effects and effectiveness of environmental policy (Vaz *et al.*, 2001). This framework was developed to assist policy makers in the EU to understand the relationship between a policy measure and its ultimate impact, especially on human behaviour and the environment. The framework contains a number of key elements: (objectives, inputs, outputs, outcomes, and impacts) that are used to evaluate a policy intervention (by the state) (Figure 2.3).

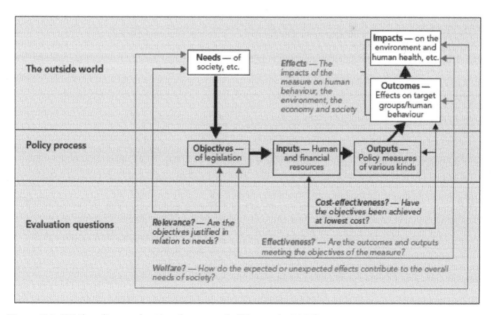

*Figure 2.3. EEA's policy evaluation framework (Vaz* et al., *2001).*

Policy processes normally involve the enactment of a regulatory framework or a piece of legislation to facilitate the implementation of a given policy programme. The relevant piece of legislation will contain the overall (policy) *objectives* of the programme and will guide the ways in which inputs, outputs, outcomes and impacts are deployed and achieved. *Inputs* refer to the number and character of resources allocated in designing, formulating and implementing the policy. These might involve resources in the form of financial resources, staffs, training, administrative structures, etc. As in any other production process, these inputs are used to produce outputs. In policy formulation, policy *outputs* are the tangible results of policy intervention. *Outcomes* are changes in the behaviour of target groups that result from policy implementation. As described by Dunn (2004), target groups are the individuals, communities or organizations whom the policy is designed to have an effect upon. The *impacts* of policy intervention or measures are the changes in the behaviour (of target groups), and on the environment. An example on the improvement of water treatment facilities can illustrate these concepts. The objective is to have an adequate supply of good quality water. The inputs will be the financial and human resource investments needed to build the treatment plants (and regulate them). The outputs will be the number (and capacity) of treatment plants built. The outcomes will be the amount (and quality) of water treated. The impacts will include an improvement in water quality for a given number of people.

In defining policy interventions or measures, one of the most fundamental issues that policy makers need to ask is who the intended beneficiaries are. What are their needs? Only by identifying and understanding the needs of stakeholders can policy makers formulate appropriate policies. The next steps are translating those needs into policy objectives. At this stage, other elements of the policy process (inputs, outcomes, and impacts) are involved. As illustrated (Figure 2.3), we can relate information from each category to the others, allowing us to make various evaluations on the implementation of a particular measure. For example, by comparing objectives with outputs and outcomes, we can say something about the *effectiveness* of a given policy – whether the goals or objectives of an intervention are being met (Vaz *et al.*, 2001). Obviously not every policy measure will achieve all its intended goals or objectives. According to Dunn (2004), policy evaluation plays an essential role in policy analysis and helping policy makers to re-formulate policy problems, re-examine the appropriateness of policy goals and objectives, and in the event of policy failure, recommend a new policy.

### 2.4.6 The American Water Works Association's QualServe business model

The QualServe Business Model – developed by the American Water Works Association – is a tool used to measure the performance of the water and wastewater utility sectors (AWWA, 2004). This model was first developed in North America and widely used there among water utilities. It has subsequently been replicated by other water utilities around the world. In developing this model the AWWA Board realized the importance of benchmarking in helping utilities improve their performance. Generally, it is based on system perspectives and used to measure the performance of water and wastewater utilities in five areas: organizational development, customer relations, business planning and management, water operation, and wastewater operations (Figure 2.4). Relevant performance indicators are assigned to each category (see AWWA 2004 for details).

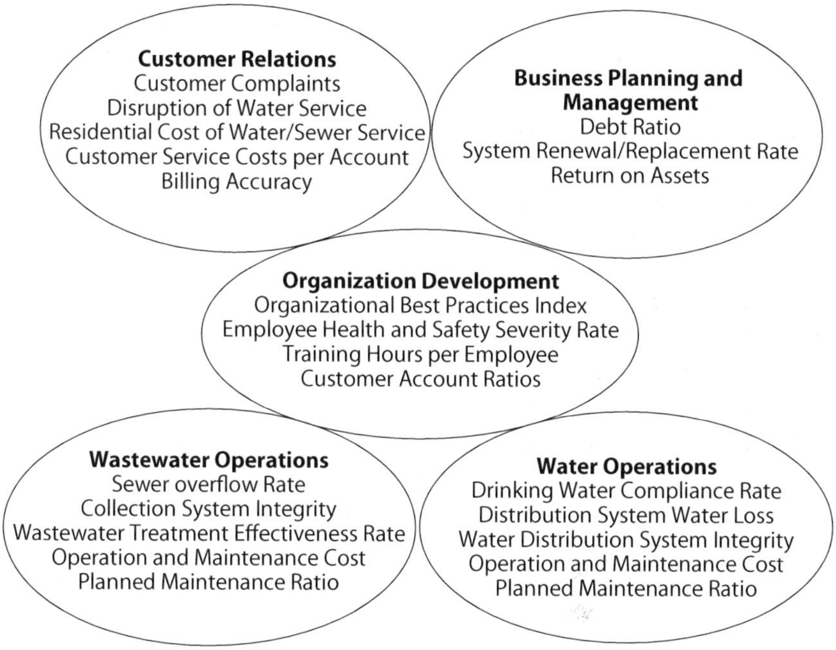

*Figure 2.4. AWWA QualServe Business Model (AWWA, 2004).*

The QualServe Business model is a useful tool for measuring (almost every aspect) of the performance of water utilities. It was designed to be used by well-developed and stable water utilities (first in North America). Thus, its application for water utilities in less developed countries might not be effective, without taking into consideration local conditions. For instance, the criteria of water distribution system integrity might not mean anything if a water utility (in developing countries) is still struggling to meet citizens' basic need for water. Equally, calculating the return on assets may not be relevant when there is no budget available for developing water assets. Of the indicators (spread across five categories) used to measure the performance of water utilities in this model, three indicators seemed very relevant to the Malaysia water supply sector. These are customer complaints, billing accuracy and the drinking water quality compliance rate. The first two are measures of operational efficiency and the third indicator refers to environmental effectiveness. Other indicators – non-revenue water, unit production cost, sludge management and information disclosure – are adopted from VEWIN and the Malaysia Water Association model (see Section 2.4.7).

This model does not provide indicators for measuring (water) sludge management, an important aspect of environmental effectiveness. This missing indicator is complemented by adopting indicators for residue[6] management (residue recycles initiative and new use of recycled residue), developed by the Association of Dutch Water Companies (VEWIN, 2006). The water utilities' performance on this indicator is measured against compliance with the Malaysian Environmental

---

[6] Sludge is the by-product of raw water treatment processes.

Quality Act 1974, and the measures or tools used by water operators to improve their sludge management. The other main environmental effectiveness indicator missing from the AWWA model and included in the present study is information disclosure.

### 2.4.7 The Malaysian Water Association's model

The Malaysian Water Association (MWA) has developed performance indicators for the water supply and wastewater sector in Malaysia. These indicators can broadly be grouped into four categories: physical performance; operational performance; service performance; and financial performance (MWA, 2006). Each category contains a number of indicators (Figure 2.5). Due to time and resource constraints experienced by the researcher only two of the MWA's indicators have been adopted: non-revenue water and unit production cost. Both indicators are used to measure the operational efficiency.

   Both the AWWA and MWA models provide indicators for evaluating the outcomes of the reform. They look at operational efficiency and environmental effectiveness. When used in combination with the policy arrangement approach, they offer a useful evaluation methodology to address two main aspects of the research questions (explained in Section 2.5). The first aspect involves evaluating the reform from the perspective of outcomes. The second aspect seeks to understand and explain the process of water supply, focusing especially on the institutional arrangements.

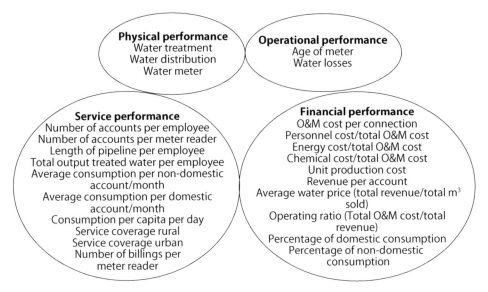

*Figure 2.5. The Malaysian Water Association Model (MWA, 2006).*

## 2.5 Conceptual framework and answering research questions

The conceptual framework of this research can now be put together by combining two main components. The first component – the policy arrangement approach – offers an analytical tool to analyse and understand the policy process of the reform, by examining four policy dimensions. The second component – the policy evaluation models – can be used to assess the outcomes of water supply reform and the operational efficiency and environmental effectiveness of the water operators. Figure 2.6 presents a schematic overview of the overall conceptual framework used in this research and Table 2.2 further relates the different elements of the conceptual framework to the current water reform research.

Sections 2.5.1-2.5.3 explain in more detail how the conceptual framework is applied to answering the research questions. To reiterate, this research addresses three main research questions. The first considers the policy process of water supply reform as public policy intervention by the state from the institutional arrangement perspective. The second question considers the contributions of the created output – the output efficacy analysis – in realising the reform's objectives. The third and last question deals with the impacts of the reform on the performance of water utilities on two areas: operational efficiency and environmental effectiveness.

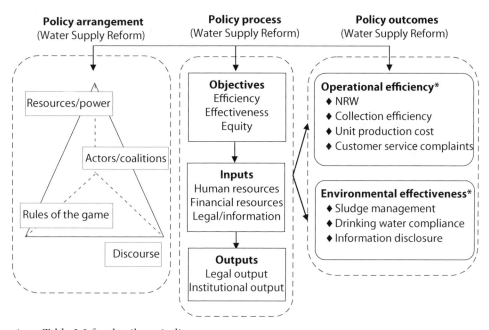

* see Table 2.3 for details on indicators

*Figure 2.6. Conceptual framework of the research (adapted from European Environment Agency's policy evaluation model and policy arrangement approach).*

*Table 2.2. Applying the conceptual framework to the research.*

| Concepts | | Applications |
|---|---|---|
| Policy arrangement approach | | Explain and understand the reform process: the objectives, actors/coalitions, resources, power, rules and discourses |
| Policy evaluation approach | European Environment Agency | Offers an evaluation framework for analysing the impacts of the reform (outcomes evaluation) |
| | American Water Works Association | Offers indicators for measuring the outcomes of the reform (operational efficiency and environmental effectiveness) |
| | Malaysian Water Association | |

### 2.5.1 Understanding and explaining the policy process of water supply reform

This question is formulated specifically to examine the overall policy process of water supply reform as policy intervention by the central government to improve the water supply sector in the country. It is answered through analytical research, where the four dimensions of policy arrangements are used to explain (and identify) actors/coalitions, resources and power, rules and discourses in order to understand the process of water supply reform in Malaysia. In particular, it analyses the differences in the policy arrangements of the reform in terms of different divisions of power among stakeholders, differences in the rules and routines of the game and changes in the dominant discourse.

### 2.5.2 Assessing output effectivenessof the water supply reform

This question examines four (new)institutions and institutional arrangements which emerged from the reform – the regulatory body, the water corporations, the financier and the water resources regulator – using policy evaluation approach as the analytical tool. As the main tenets of the reform, how effective these institutions can contribute (will) determine the realization of the reform's objective. As such, analysing output effectiveness allows policy makers to formulate or suggest policy interventions (in case of ineffectiveness), or strive for (better) performance (in case of effectiveness) which eventually lead to the improvement of the reform process.

### 2.5.3 Measuring operational efficiency and environmental effectiveness

This third question seeks to analyse the impacts of the reform on operational efficiency and environmental effectiveness. For this purpose, an analysis will be conducted on a set of chosen indicators assigned to both (Table 2.3). For operational efficiency the following indicators will be used: non-revenue water, collection rate efficiency, unit production cost, and customer

*Table 2.3. Operationalization of operational efficiency and environmental effectiveness indicators.*

| Indicator | Mathematical formula |
|---|---|
| **Operational efficiency** | |
| NRW | $\dfrac{\text{Amount of water supplied} - \text{amount of water consumed/billed}}{\text{amount of water supplied}} \times 100$ |
| Collection rate efficiency | $\dfrac{\text{No. of bills/revenue issued during reporting period}}{\text{no. of bills/revenue collected during reporting period}} \times 100$ |
| Unit production cost | Cost to produce water / amount of water produced (in million litres) |
| Customer service complaints | Per 1000 connections/accounts |
| **Environmental effectiveness** | |
| Sludge management | • The availability of on-site sludge treatment facilities <br> • Compliance to EQA 1974 in terms of sludge management <br> • Adoption of environmental-friendly sludge treatment facilities <br> • Sludge recycling and re-utilization <br> • Environmental concerns: <br>     • environmental pledge (policy statement, pledge, charter); and <br>     • adoption of green tax |
| Compliance with drinking water standard | • The state of the compliance with NGDWS 2001 <br> • Readiness to comply to Section 41 of WSIA |
| Information disclosure | • Information disclosure on sludge management and compliance with drinking water quality standards <br> • Readiness to comply to Section 29 of WSIA |

service complaints. For environmental effectiveness the following indicators will be used: sludge management, compliance with drinking water quality standards and information disclosure.

From the policy intervention perspective, analysing the impact of the reform should also entail evaluating the policy intervention against its set goals or objectives. Here the policy evaluation approach provides a framework for evaluating the outcomes of the reform on operational efficiency and environmental effectiveness. Policy evaluation can also be used to identify other factors that might influence the performance of water operators on operational efficiency and environmental effectiveness. By analysing the impacts (and outcomes) of the reform we can make an overall assessment of the extent to which the objectives of the reform were met. In the event that objectives were not met, or only partially attained, policy makers can adopt modifications to the policy. Within the policy evaluation framework, the performance indicators developed by AWWA and the MWA are used to measure the performance of water operators on operational efficiency and environmental effectiveness for a period before and after the reform.

# Chapter 3.
# The water supply reform process in Malaysia

*'The reform model that we are embarking on is unique and I hope it will serve as a guide to developing as well as developed countries'*

– a foreword by the Honorable Minister of Energy, Water and Communications in *The Water Tablet: Malaysian Water Reform*, 2008.

## 3.1 Introduction

The above quote indicates the pride the Malaysian government has taken in the reform of the water sector so far. The first step in the current water sector reform was taken by re-visiting the policy decision of the National Water Resources Council (NWRC) meeting in 2003 to restructure the water sector by bringing all water management under the responsibility of the federal government. Just after its establishment in 2004 as the first ever dedicated 'water ministry', the Ministry of Energy, Water and Communications (MEWC) presented a new reform model to the government. This new model was based on the premise of shared responsibility between the federal and state governments over water management which recognized the jurisdiction of state governments over water resources and conferred power to the federal government on regulatory matters related to 'water supply and services'. It was motivated by the recognition that reform was needed to bring about a water supply sector that was sustainable in terms of efficient operation, effective regulation and feasible financial mechanisms. Two sections in this chapter sum-up the historical process of the reform. Section 3.2 presents the events leading to the reform and Section 3.3 looks at the driving forces that drove the reform process. Chapter 1 noted that the reform resulted in the establishment of several institutions but did not look at how they contributed to the attainment of the reform's objectives. This question is examined in the second part of this chapter: analyzing the outputs of the water reform and the extent to which the water reform has achieved its intended objectives (in terms of concrete results). It does so by comparing the results of the reform (in terms of laws and institutions) to its objectives. After identifying and defining the policy objectives of the reform (Section 3.4), an output efficacy analysis is conducted (Section 3.5). Section 3.6 provides the overall conclusion of this chapter.

## 3.2 The events leading up to the reform

The history of the water sector reform in Malaysia unfolded through four important events. It started with the policy decision taken during the NWRC meeting in 2003 to hand over all water management to the federal state. The formation of the MEWC after the 2004 general election accelerated the implementation of this decision, but in a revised form: that of shared responsibility between states and the federal state. The third event was the commissioning of a feasibility study to identify the best possible option to implement that decision. Fourth and lastly, the Federal

Constitution was amended (in 2005) allowing the entire reform process to take place. This section follows the historical track of these four events.

### 3.2.1 *The decision of the National Water Resources Council*

Before 2004, water supply was under the Ministry of Works (MOW); the main ministry responsible for the national infrastructure development programmes. Its technical arm, the Public Works Department was mandated to carry out water supply works, in close collaboration with state governments. The National Water Resources Council (NWRC) was established in 1998 to administer water resources. Its mandate was 'to pursue more effective water management, including the implementation of inter-state water transfers' (Zaharaton, 2004: 3). In 2003, the MOW tabled the proposal in a NWRC meeting[7] for the federal government to take over water management from the state government by means of amending the Federal Constitution (MEWC, 2008). The MOW believed that a constitution amendment was needed for the state government to relinquish power over water to the federal government. The Chief Ministers (of state governments) were then expected to brief their respective (state governments) Rulers on the decision of the meeting, but this did not happen. As such, the decision was not implemented and the constitution was not amended. Hence, the *status quo* in water management remained. Nevertheless, the MOW had set the first move which eventually led to realization of the reform in 2004. In 2004 the first ever dedicated water ministry, the MEWC was formed. It was then that the idea for reform was revived.

### 3.2.2 *After the 2004 general election*

The 11th general election, held in February 2004, was significant to the water supply sector in the country as it led to the creation of the MEWC which took over responsibility for the water supply function from the Water Supply Division[8] and waste water function from the Sewerage Services Department.[9] Under the leadership of the (then) Honorable Minister Lim Keng Yaik, the NWRC's earlier decision to federalize water management was thoroughly reviewed. Finally, after series of deliberations a 'new model' of reform was decided upon. This model was a departure from the initial proposal as it acknowledged the sovereignty of state governments over water resources, and the need for federal intervention on regulatory oversight and financial mechanisms. In other words, water management was to become a shared responsibility between the federal and state governments. However, the MEWC had limited experience on water reform. They needed to develop a workable model which could address the operational aspect of the sector and the economic aspects of water supply such as tariffs, regulation and financial mechanisms. It was obvious that it did not have sufficient resources to do this itself. This led the MEWC to commission a consortium of consultancy firms to assist in drafting a model for reform. In 2004, the formal commissioning of a feasibility study was finalized.

---

[7] Chaired by Prime Minister and attended by all Federal Cabinet Ministers and State's Chief Ministers.

[8] Previously under the Ministry of Works.

[9] Previously under the Ministry of Housing and Local Government.

### 3.2.3 Commissioning the study

Taking the shared-responsibility principle as a premise, the reform aimed to address three main fundamental aspects of the water supply sector. The first related to improving the operational efficiency of the sector. The second addressed the absence of effective regulation by establishing a central regulatory regime. The third concerned the financial mechanism(s) needed to address over-dependency on federal budgets and expensive private borrowing. Given the broad range of issues that the reform sought to address, a consortium of consultancy experts with operational/technical, economic and legal backgrounds was appointed in August 2004. The main anchor firm was KPMG, assisted by a legal firm Zulrafique and SMHB, an engineering consultancy. The assignment had two principle objectives: (1) to propose a viable and economic structure for the water supply sector in the country; and (2) to facilitate the federal government in putting in place a policy and regulatory framework for the orderly and sustainable development of the water supply sector (MEWC, 2004).

Besides relying purely on external consultants, an attempt was made to learn from the success stories of other countries. A high ranking study tour, led by the Minister, went to France and the UK. Preliminary findings from the tour suggested that an (independent) regulatory regime, following the UK model, was seen as appropriate to the water supply sector in Malaysia.

After having considered the findings from the tour, the preliminary report of the study was presented to the government. One of the key proposals was the formulation of a legal and regulatory framework for the sector (see Section 3.4) (MEWC, 2008). In a nutshell, the consultants concluded that the water supply sector urgently needed to address two fundamental issues: efficiency and effectiveness, and financial constraints (see Section 3.3).

Commissioning the study was not enough to start the reform. The federal government could only become involved in water management when amendments to the Federal Constitution were made. These amendments were to be done in line with the shared-responsibility principle discussed in Section 3.2.2.

### 3.2.4 Amending the Federal Constitution

Under the federal system of government, there is a clear division of powers between the federal and the state governments. The Ninth Schedule of the Federal Constitution clearly divides the legislative jurisdiction into three categories, namely:
1. State List for matters exclusively under the jurisdiction of state governments;
2. Federal List for matters exclusively under the jurisdiction of the Federal government; and
3. Concurrent List for matters where both Federal and state governments have powers to legislate.

One of the matters under the State List was water (including water supplies, rivers and canals) (MEWC, 2008). This gave the power to legislate on matters concerning water exclusively to state governments. This meant that each state had its own set of legislation pertaining to water – the State Water Enactment – only applicable to that particulate state. Thus to facilitate the reform, the Ninth Schedule of the Federal Constitution first needed to be amended. This involved moving 'water matters' from the State List to the Concurrent List. The MEWC was charged to work closely with the Attorney General's Chamber to affect these amendments, which were eventually completed in early 2005 (MEWC, 2008).

The amendments gave federal government control over water from the point of 'abstraction of raw water' to the point of 'supply to consumers' (MEWC, 2008). This left state sovereignty over water resources (including land) and ownership over water utilities unchanged. Section 3.4 discusses how a central regulatory body was created to take over the role previously played by state governments.

## 3.3 The driving forces behind the water sector reform

As highlighted in Section 3.2.3, the reform was initiated to address two fundamental problems faced by the water supply sector: efficiency and effectiveness, and funding constraints.

### 3.3.1 Improving efficiency and effectiveness

In general, water utilities across Malaysia could be described as being inefficient and ineffective in several respects: they had high levels of non-revenue water, low revenues, unsustainable tariff structures and were weakly regulated by the state regulators.

### High levels of non-revenue water

Non-revenue water refers to the difference between the amount of water supplied and amount of water billed or metered. That difference is the amount of water which is lost or which enters the distribution system but does not bring any revenue to the water utilities (Winarni, 2009). According to the Infrastructure Leakage Index, developed by the International Water Association (IWA), there are three categories of (water) losses: real/physical losses, apparent losses and unbilled authorized consumption (Figure 3.1). Real losses are mainly losses which occur due to leakages. Apparent losses are the result of unauthorized consumption, such as water theft and illegal connections, or metering inaccuracies. Unbilled authorized consumption refers to both unbilled metered or unmetered consumption.

High non-revenue water had been plaguing water utilities in Malaysia for decades. Thus, it came as no surprise at all that one of the key motives of the reform was to address this problem. In 2005, the nationwide average for non-revenue water stood at 38% (MWA, 2005). In some water utilities, the figure was as high as 50%. This problem of non-revenue water is widespread in Asian countries. The Asian Development Bank (2010) reported that water utilities across Asia recorded an average non-revenue water of 30%. This has negative effects on all water utilities, causing them to operate at a low level of efficiency, increasing the costs of water collection, treatment and distribution, and reducing water sales (and revenue). This in turn makes it harder for water utilities to keep their water tariffs at reasonable and affordable levels.

A high level of non-revenue water also demonstrates inefficient management of the water supply. Prolonged non-revenue water causes water utilities to lose millions in revenue, which could be used to make other service improvements. It was intended that the reform should find a remedy to this situation.

| | | Billed authorized consumption | Billed metered consumption | Revenue water |
|---|---|---|---|---|
| | Authorized consumption | | Billed unmetered consumption | |
| | | Unbilled authorized consumption | Unbilled metered consumption | |
| | | | Unbilled unmetered consumption | |
| System input volume | | Apparent losses | Unauthorized consumption | Non-revenue water |
| | | | Metering inaccuracies | |
| | Water losses | Real losses | Leakage on transmission and/or distribution mains | |
| | | | Leakage and overflows at utilities' storage tanks | |
| | | | Leakage on service connections up to point of customer metering | |

*Figure 3.1. IWA's Infrastructure Leakage Index (ILI) (Farley, Wyeth, Md. Ghazali & Singh, 2008).*

## Unsustainable water tariff structures

The reform was also implemented to correct unsustainable water tariff structures, which were the result of two main factors. The first was that water tariffs did not reflect the true cost of abstracting, purifying, distributing water and treating waste water – which was (highly) subsidized. Purification and distribution costs usually represent the largest components of water tariffs. Most (if not all) state governments did not impose royalties for raw water abstraction, which was rarely captured in the tariff structure. Consumers paid their waste water charges separately. Highly subsidized water tariffs (intended to increase access and equity) caused the water utilities to face financial deficits. A second reason was that certain water utilities enjoyed or had exclusive rights in terms of a guaranteed tariff revision and/or guaranteed returns for their concessions, which had enormous impacts on the tariff setting mechanisms (MWA, 2008a). The following paragraphs explore how these factors led to unsustainable tariff structures.

Table 3.1 shows the tariffs of some Malaysian water supply entities in 2005, where 75% – all of which were state-run water utilities at the time – recorded the tariffs below the national average. These state entities (and their political masters) considered water to be an essential element in promoting people's health and livelihoods and aimed to ensure that citizens' water needs were made at the lowest possible cost. In most cases, subsidies were used to keep water tariffs affordable. The lowest tariff was recorded in the state of Penang. Elsewhere the state government of Selangor was subsidizing every cubic meter by RM 0.60, to keep the tariff affordable. Such below-cost tariffs not only hurt the financial standing of water utilities, but were also detrimental to the environment. Cheap water led to over-consumption which eventually degraded the environment through excessive waste water discharges. It also meant that most state-owned water utilities could not generate enough revenue to re-coup their investments or to undertake new ones.

Imposing cost-reflective tariffs (or at least a tariff increase) was politically difficult. Such decisions were unpopular with state governments fearing they might backfire and lose political support (and power). As a result, tariff revisions were delayed. In some states, like Sabah and

Table 3.1. Domestic and industry water tariffs in 2005 (RM/m³) (MWA, 2005).

| Water supply entities | Supply areas (state) | Domestic tariff for first 35 m³ | Industry tariff for first 500 m³ | Average tariff |
|---|---|---|---|---|
| PBAPP | Penang | 0.31 | 0.94 | 0.63 |
| SATU | Terengganu | 0.52 | 1.15 | 0.84 |
| JBA Kedah | Kedah | 0.53 | 1.20 | 0.87 |
| AKSB | Kelantan | 0.55 | 1.25 | 0.90 |
| PWD Perlis | Perlis | 0.57 | 1.30 | 0.94 |
| JBA Pahang | Pahang | 0.57 | 1.45 | 1.01 |
| LAP | Perak | 0.67 | 1.40 | 1.04 |
| JBANS | N. Sembilan | 0.68 | 1.59 | 1.14 |
| Syabas | Selangor[1] | 0.72 | 1.91 | 1.32 |
| PAM | Melaka | 0.72 | 1.40 | 1.06 |
| JBA Labuan | Labuan | 0.90 | 0.90 | 0.90 |
| SAJH | Johor | 0.90 | 2.93 | 1.92 |
| Nat. average | | 0.64 | 1.05 | 1.04 |

[1] Include Kuala Lumpur and Putrajaya.

Labuan, tariffs had not been reviewed since 1982 (MWA, 2008a). One of the intentions of the reform was to depoliticize decisions on tariff increases. Under the reform, decisions on tariffs would be decided by the central regulator, thus diminishing the influence of state politicians. In other words, centralized decision making over tariffs would create the opportunity for a gradual implementation of a cost-reflective tariff structure, while balancing this against the interests of poorer sections of society through a policy which continues to subsidize the 'lifeline consumption' (of up to 30 m³ per month/household) or through a direct subsidy scheme.

Under privatization concessions, water operators were guaranteed a tariff increase in exchange for them making investments in the sector. In Selangor, Syabas[10] was allowed a tariff revision every three years, while the internal rate of return for SAJH[11] was fixed at between 14% and 18%. These arrangements did not work in favor of state governments. On the one hand, raising tariffs could have caused them to lose political power when the next elections came around. On the other hand, refusing a tariff increase could have caused them to have to pay millions in compensation to private sector. Weak state regulations also hindered reform as state governments often faced a conflict of interests, being regulators themselves but also involved in the water business. The matter became even more complicated when issues of 'political incompatibilities' between central and state government arose: when federal and state governments were governed by two different

---

[10] Private water company responsible for supply in Selangor, the Federal Territory of Kuala Lumpur and Putrajaya.

[11] Private water company in Johor.

political parties. This happened in Selangor, where the new state government refused to grant a tariff increase to a private water operator as it suspected that the deal (agreed by the previous government) overly favoured the private utility. The matter was finally settled by an intervention from the central government, which ensured that RM 600 million was paid to the private operator for delaying the tariff increase.

## Weak state regulation

Regulation of the water supply sector was decentralized, with state governments regulating their water sector individually. There was no single central regulator. Through their respective water departments state governments were self-regulating, and often this meant no effective regulation existed at all. This type of self-regulation usually existed in states where water departments remained a government department, such as in the states of Kedah, Perlis, Labuan and Pahang. In such situations self-regulation has little effect on the performance of state water departments, due to an absence of effective performance measures and, to some extent, conflicts of interests. In one interview, the Chief Executive Officer of a private water operator[12] believed that conflicts of interests were common when state governments performed the role of 'hunter and poacher' at the same time.

In states where water supply was corporatized or privatized, state regulation was also ineffective in regulating the behaviour of water companies. Normally this body was established within the existing state apparatus and controlled by the state administration. Such arrangements existed in Kelantan, Terengganu, Penang, Johor and Selangor. These were not effective for two reasons (MEWC, 2008). First, as part of the state administration (where the Chief Minister is the head of the state) state regulators were merely public servants with limited powers to undertake effective regulation. Second, conflicts of interests occurred through the involvement of state governments in the water business (through joint-ventures with the private sector). For example, the state government of Selangor had business interests in all four of the private water companies working in the state. In both circumstances, the state regulator felt helpless to deal with the (politically well-connected) water companies (MEWC, 2008).

Hence, neither self-regulation nor state regulation was able to increase the efficiency of the water supply sector. Both forms of regulation are weakened by conflicts of interests resulting from the intermingling of service provision and regulation and the business interests of state governments. The reform sought to address these shortcomings by establishing an effective and independent central regulator which would help improve the performance of the sector. One way in which efficiency gains could be achieved was through a clear separation of power between owners, service providers and regulators. Such a separation of power helps to reduce conflicts of interests and to minimize political interference on tariffs. In addition, the reform was designed to enable the implementation of a standardized set of performance indicators for all water operators, a key step towards benchmarking or comparative regulation.

Thus, one of the main goals of the reform was to establish a strong and largely independent central regulator to regulate the behaviour of water operators. This central regulator would initiate

---

[12] Personal interview on 30 Apr. 2009.

transparent and well-defined standardized performance indicators for all water utilities and promote their long-term efficiency.

### 3.3.2 Bridging funding gaps

*Financial model*

The water supply sector was previously financed by two main sources: public funding[13] and private funding. Public funding was mainly provided by the federal government to state governments, which channeled funds to their respective state water departments. These funds were provided as interest-free loans. Meanwhile private companies mostly relied on private funding to fulfill their contractual obligations under water privatization. Both sources have proven to be unsuccessful in facilitating the long term goal of promoting efficiency in the sector.

Even though public funding for water increased every five years (Table 3.2), it has not matched the requirements of state governments. Furthermore, public funding was highly contested by the other sectors as well. Over-dependency on limited public funding affected the ability of state governments to improve the water supply system. Their inefficiency to manage non-revenue water, coupled with below cost water tariffs, placed most state governments in a deficit cash flow position, meaning that they could not re-pay the federal loans. As of 2005, there was RM 7.6 billion of loans outstanding from state governments (MEWC, 2008). This prevented the water sector from improving its performance or growing, as insufficient investment was available to develop the sector. It was obvious that public funding could not provide the massive investment needed by the water sector, especially in view of projections about future economic growth.

As public funding was not available to the private water sector, they relied on private funding to fund the capital expenditure works agreed as part of the privatization concessions with state governments. For instance, in 2005 Syabas had to raise RM 7.1 billion for capital expenditure to undertake the privatization of the water sector in Selangor. Such funding arrangements exposed private water operators to the financial risk of funding long-term water projects with short-term loans. This arrangement put tremendous pressure on water tariffs as private water utilities were forced to recoup their investments as well as make profits for their shareholders.

For the private sector raising tariffs was the most feasible way to recoup their investments. For instance, under the 2005 privatization, water tariffs in Selangor were allowed to be reviewed every three years (Syabas, 2005). The first revision occurred in 2006 with a water tariff increase of 15%. The second revision was due in 2009 with a larger increase of 37%. This phenomenon of guaranteed tariff revision was not a good thing for the water sector as it did not encourage long term efficiency gains in the sector. The reform was introduced to enforce performance-based regulations based on a 'carrot and stick' approach. The reforms aimed to address both these problems, reducing constraints on public funding and cushioning excessive tariff increases caused by private funding.

---

[13] Includes borrowing from international organizations such as the World Bank, ADB, etc.

## Federal cash flows to the water sector

Water utilities in Malaysia were predominantly state-owned (public utilities), usually taking the form of state water departments or state water corporations. In some cases state governments also had stakes in private companies. Federal-state relationships duty bound the federal government to assist state governments with developing national utilities, including water supply. It was unusual for the federal government to give loans directly to water utilities. Record showed that the first federal financial assistance (of RM 538 million) was given under the Third Malaysia Plan (1976-1980) (Economic Planning Unit, 2008). As national income increased, federal budgets for water infrastructure development followed suit in subsequent five year plans (Table 3.2).

However, these increases fell far short of the projected budgets for the future development of water projects. The National Water Resources Plan (2010-2050) reported that RM 22.2 billion was needed over the period 2000-2010 to ensure that Malaysia had enough clean and safe drinking water. This was based on the estimation that water demand rose by 6% during this period (Economic Planning Unit, 2000, 2008).

Thus the available financial resources were not sufficient to meet the investment requirements for water infrastructure works. Continuous dependence on federal assistance was delaying the necessary water infrastructure development, which would eventually lead to problems with supply and water quality. This forced the federal government to explore alternative financial resources. One aim of the reform was to tap into financial resources available outside the public domain. A single purpose government-owned company, the Pengurusan Aset Air Berhad (PAAB), was established in 2006 to bridge the funding requirements for water infrastructure development (see Section 3.5.3).

## 3.4 Analysis of the outputs of water reform

In this section, we first identify and define the policy objectives of the reform. This is followed by identifying the resultant policy output.

Table 3.2. Water infrastructure allocations under successive Malaysia Plans (Economic Planning Unit, 2008).

| Malaysia Plan | Period | Total allocation (RM million) |
|---|---|---|
| Third Malaysia Plan | 1976-1980 | 538 |
| Fourth Malaysia Plan | 1981-1985 | 2,085 (+287%) |
| Fifth Malaysia Plan | 1986-1990 | 2,348 (+13%) |
| Sixth Malaysia Plan | 1991-1995 | 2,089 (-11%) |
| Seventh Malaysia Plan | 1996-2000 | 2,385 (+14%) |
| Eighth Malaysia Plan | 2001-2005 | 4,000 (+68%) |

### 3.4.1 Identifying and defining policy objectives

The objectives of water sector reform in Malaysia can be found most explicitly and clearly in the Cabinet Paper – a policy paper presented to the Cabinet by the MEWC in 2005. As explained above, the objectives of the reform were derived from a study commissioned by the MEWC, following the government decision to restructure the water sector. These objectives were then translated and operationalized into a legal output known as the Water Services Industry Act (WSIA) (see Section 3.4.2 and Box 3.1). In principle, the prime objective of the reform (as encapsulated in the WSIA) was to put the water sector on a sustainable footing. This (as previously discussed) involved addressing two fundamental issues: funding constraints in the water supply sector and a more effective and efficient performance by the water supply sector as a whole (MEWC, 2008). Meeting these objectives and resolving these two fundamental issues involved creating a number of outputs, which will be summarized in the next Section. After this, the outputs will be analyzed with respect to four goals (described in Section 3.5): (1) regulation; (2) the organization of water resource management; (3) financial institutions; and (4) operational management.

---

**Box 3.1. Ten national policy objectives for water supply and sewerage services industry (Water Services Industry Act 2006).**

Policy objectives of the WSIA:
a.  to establish a transparent and integrated structure for water supply (and sewerage) services that delivers an effective and efficient service to consumers;
b.  to ensure the long term availability and sustainability of the water supply including the conservation of water;
c.  to contribute to the sustainability of water courses and water catchment areas;
d.  to facilitate the development of competition in the industry to promote economies and efficiency in the water supply and sewerage services industry;
e.  to establish a regulatory environment that facilitates financial self-sustainability amongst industry players in the long term;
f.  to regulate for the long-term benefit of consumers;
g.  to regulate tariffs and ensure the provision of affordable services on an equitable basis;
h.  to improve the quality of life and environment through effective and efficient management of water supply and sewerage services;
i.  to establish an effective system of accountability and governance between industry players; and
j.  to regulate the safety and security of the water supply and sewerage systems.

---

### 3.4.2 Policy outputs

The tangible outputs of the reform can be divided into two inter-related groups: legal outputs and institutional outputs.

### Legal outputs

The two main legal outputs of the water sector reform process are the WSIA and the *Suruhanjaya Perkhidmatan Air Negara* Act (SPANA) or the National Water Services Commission Act. WSIA addresses most of the issues related to efficient and effective operations (water supply, water quality, licensing, business approach, etc.), while SPANA deals with independent and effective regulation of the water sector, through the formation of a central regulation authority, the National Water Services Commission (NWSC).

WSIA was enacted with the main objective of regulating water supply services (MEWC, 2008). The central tenet of WSIA was to establish a licensing and regulatory framework to uphold the national policy objectives for sustainable water supply (and sewerage) services. In this respect, WSIA supports the national policy objectives by introducing a standardized licensing and regulatory framework imposed and enforced upon all water utilities. Previously, this function was administered independently by state governments. Standardized regulation is to be achieved through introducing a uniform licensing requirement and regulatory oversight. WSIA introduced two types of licenses. An *individual license* is required for water utilities operating a public water supply system, while a *class license* is needed to operate a private water supply system (i.e. in a plantation/estate or in remote areas). All water utilities (public or private) are also required to have a *facilities license* if they own a water supply system.[14] Thus, the WSIA consolidates the operation of all water utilities under one single regulatory body, the NWSC.

However, the enforcement of WSIA does not affect the general application of existing laws on environmental quality, land matters or state governments' existing powers over water resources. This means that the Environmental Quality Act 1974 remains the main legal framework for regulating environmental matters; and land and water resources remain under the control of state governments (as guaranteed under the Ninth Schedule of the Federal Constitution) (MEWC, 2008). Here the reform sets a clear demarcation between the functions, tasks and powers that fall under the jurisdiction of the federal government and those that fall under state governments.

SPANA spells out the mechanisms through which the central regulator, the NWSC, operates. It deals mainly with the formation of the NWSC, its membership, powers and functions, employment and financial matters (MEWC, 2006b). In a nutshell, SPANA established and set the terms and goals of the central regulator, the NSWC. It clearly defines the NWSC's responsibilities for water supply and services, with regulation of water resources falling outside of its jurisdiction. State

---

[14] The WSIA defines the water supply system as 'the whole of a system incorporating public mains, pipes, chambers, treatment plants, pumping stations, service or balancing reservoirs or any combination thereof and all other structures, installations, buildings, equipment and appurtenances used and the lands where the same are located for the storage, abstraction, collection, conveyance, treatment, distribution and supply of water' (MEWC, 2006a, pp. 18).

legislations needed to be amended or enacted (by the state governments) to provide legal back-up for the management of water resources. This usually involved state governments establishing a water resources regulator.

*Institutional outputs*

Institutional outputs of the water sector reform are the NWSC, the state water corporations, the state water resource regulators and PAAB. Both the NWSC and PAAB operate at the federal level (as federal agencies); whereas water corporations and state water resources regulators are state agencies operating at the state level.

The first output – the NWSC – was officially established in 2007 to enforce the WSIA and to drive efficiency and effectiveness gains in the water sector. The NWSC's main functions can be summarized into three key areas: (1) implement and promote the national policy objectives for water supply and sewerage services; (2) promote a fair and efficient mechanism for determining tariffs and implement tariffs that have been established through appropriate mechanisms and tools; and (3) undertake operational activities relating to non-revenue water, supply, coverage, access and quality (MEWC, 2006b).[15]

Section 5 of SPANA defines who can be appointed members of the Board of the NWSC, which consists of a Chairman, a Chief Executive Officer (CEO) and not more than 10 other members (MEWC, 2006b). All of them are appointed by the Minister. With the exception of the CEO, the members of the Board of the NWSC can hold office for a term not exceeding 10 years. In addition, the NWSC may 'establish any committee as it considers necessary or expedient to assist in the performance of its functions' (MEWC, 2006b: 12). At the working level, NWSC is headed by the CEO and supported by several departments covering (for example) the water supply and sewerage sectors (see Appendix 2).

The second output is state water corporations. These were established to take over the roles of the (now defunct) state water departments. As one of the main pillars of the water sector, state water corporations are expected to improve their efficiency in both operational terms (supply, access, quality, etc.) and in the economic domain (e.g. profits, returns on investment, collection efficiency, etc.). These goals were to be achieved by injecting private sector cultures. In addition, state water corporations facilitate (the implementation of) licensing requirements. Under the reform, a license can only be granted to a business entity, without which standardized licencing regulations can not be effectively enforced. The latest figures show that five water utilities – two private companies (SAJH, PBAPP) and three water corporations (SAMB, SAINS, SAP) – have been granted a license under WSIA (SPAN, 2011b). However, the success rates in converting state water departments into water corporations vary between state governments. Here several critical issues emerge. What happens in the interim period when the corporatization plans are still being worked out? Do water utilities need to cease operation while waiting for a license to be granted, or risk a penalty by illegally providing water? The reform addressed this concern by suggesting a 'cooling-off' period which allows water utilities to legally provide water without a valid license from NWSC. This gives water utilities three months (for water utilities with a concession, licence

---

[15] Please refer to Section 15 of SPANA for more details.

or/and permit) and one year (for state-run water utilities) to carry out operations without a licence; although these periods can be extended by the NWSC (see Section 188 WSIA). This practice of extending running water supply without a licence is a common practice because it takes longer period to decide on corporatization by the state and federal agencies. This permission for extension has three objectives: (1) to avoid the interruption to the water supply (during the transition period); (2) to give adequate time to the state and federal agencies to take necessary steps to corporatize the water department; and (3) to facilitate the migration of private water utilities (with concession, permit or/and licence) into the new licensing regime under the reform.

The third output is the state water resource regulator. Usually this institution was established under the respective state water laws and this continues to be the case in the post-reform period. This is because state governments' kept their power over water resources under the reform. To a certain extent, the reform has re-emphasized the importance (and urgency) of having such bodies at the state level. It emphasizes the importance of protecting and conserving water resources, and the need to have a dedicated body to ensure continued availability of (raw) water in the future. State water regulators have now been established in Selangor, Johor, Penang, Kedah and several other states.

The fourth and last institutional output of the reform is the water sector financier, the PAAB. Established in 2006 PAAB's main objective is to bridge the funding gap and address the absence of a sustainable financial model in the water sector (as explained in Section 3.3.2). Its areas of responsibility include revenue and lease management, procurement, business planning and water infrastructure development (PAAB, 2010). PAAB is a wholly-owned government company but with a private sector working culture in terms of human resource and financial management and operational autonomy. It is headed by a Chairman appointed to the post by the Minister of Finance, the sole owner. A Board of Directors (of 7 members) charts the policy direction of PAAB (PAAB, 2010). These directors are drawn from both the public and private sectors. A Chief Executive Officer leads the management team which comprises several departments (see Appendix 3).

The financial mechanism adopted by PAAB works on a 'build and lease' platform (MEWC, 2008). In principle this concept gives PAAB the mandate to secure funding to develop water infrastructure and then lease this infrastructure to water utilities for an agreed lease rental. The main benefit of this approach is that it permits PAAB to consolidate the development of water infrastructures. In this way the planning, implementation and monitoring of water projects can be efficiently coordinated within one single entity.

## 3.5 Assessing the efficacy of the outputs

In this section, the effectiveness of the reform will be assessed by analyzing the outputs against the intended policy objectives (following Gysen, Bruyninckx & Bachus, 2006). This will be done by concentrating on four key aspects. With respect to regulation, I look at how effective the NWSC actually is in regulating the sector. I assess the role of state water resource regulators (or state governments) in safeguarding water resources. An assessment is made of the role that PAAB plays in stabilizing the sector's finances. And lastly, an assessment is made of whether the corporatization of the state water departments addressed the efficiency and effectiveness of those water utilities.

### 3.5.1 Regulation

The essence of the SPANA was to pave the way for the formation for the central industry regulator, the NWSC. As one of the crucial pillars of the reform, NWSC was established to support the regulatory framework needed to establish effective regulation. Modeled upon the Office of Water (OFWAT) in the UK, the NWSC is expected to set an example of good governance, separating policy making, service provision and its regulatory functions (Rouse, 2007). Industry observers have high expectation that the NWSC will be able to operate independently.

Interviews revealed a general consensus among stakeholders that the NWSC requires a certain degree of (political) 'independence' to function effectively. However, civil society organizations, such as FOMCA,[16] believed that subjecting regulatory functions to the Minister's discretion exposes the NWSC to political interference. This interference has somewhat hindered the NWSC from acting as independently as initially expected. However, it is usual practice in a political system like in Malaysia for a public body such as the NWSC to be answerable to the politician heading the relevant ministry. In the case of the reform, the NWSC was obliged to execute the policy directives set out by the MEWC. In this regard, the representatives of MEWC and the NWSC were aware that (total) freedom in regulating the water sector did not exist. A representative of the MEWC concurred that 'the NWSC is not totally free from (political) interference in executing its duties, as it is responsible to its political master, the government of the day'.[17] A representative of the NWSC said that, as much as it wants to be independent, it is 'aware that the kind of independence needed must fall within our own mold'.[18]

After having been in operation for almost 4 years, the NSWC has – to a certain extent – benefitted from the political ties it has with politicians and the government: the Minister and the MEWC. Without political support, it would have been difficult for NWSC to accelerate the corporatization agenda (with five utilities now corporatized) and tariff revisions in four states[19] (MWA, 2011). Nevertheless, as a law-enforcing agency, it is crucial that the NWSC is publicly perceived as being 'clean' and transparent in enforcing laws. It is here where its 'political ties' might work against the NWSC. Political meddling has been one of the major root causes which crippled the functions of regulatory bodies and water utilities in several countries (see Kessides, 2005; Holland, 2005). No one can guarantee that this will not happen to the NWSC even though the organization wants to send a clear signal that its decisions are made without intimidation or influence by individuals or organizations. Its Chief Executive Officer emphasized that 'we remind all our clients not to give or donate contributions (in kind or cash) in exchange for favours'.[20] A notification placed at the entrance of a NWSC office reaffirms its desire to stay impartial and not accept gifts (Figure 3.2). Nonetheless, like any organization, NWSC is not immune from interference. The danger is that ambitious politicians vying for business opportunities (in the

---

[16] Personal interview on 10 Aug. 2009.

[17] Personal interview on17 Sept. 2009.

[18] Personal interview on 15 Sept. 2009.

[19] Personal interview with MEWC representative.

[20] Personal interview on 23 Sept. 2009.

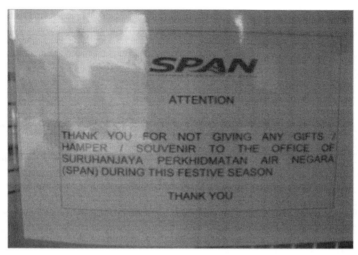

*Figure 3.2. Notice at the NWSC office refusing any gifts.*

water sector) or water utilities wanting to use their political connections may want to influence the decisions of the NWSC – even though there is no evidence to date of the presence of such influence.

Some question the impartiality of the NWSC and whether it has been granted excessive powers. Water utilities view several provisions of the WSIA as 'unfriendly'. If it is fully enforced they would have an adverse effect on water utilities in general. Section 114 has received much attention: it empowers the NWSC, on behalf of the Minister, 'to assume control of property, business and affairs of licensees in the national interest' (MEWC, 2006a: 88). They feared that this section might be abused by the Minister. They particularly question the manner in which the Minister has the discretion to decide what entails 'the national interest', and that his decision cannot be challenged in any court of law.[21]

Others see WSIA as being too punitive to water utilities.[22] Several of its provisions subject water utilities to (severe) punishment or penalties for non-compliance with regulations. They cited Section 29, on the obligatory furnishing of information, as an example. This section gives the regulator powers to ask for information from water utilities as and when required. Most water utilities, especially public water departments, will find this provision hard to comply with, making them liable to a hefty fine of up to RM 20,000 (MEWC, 2006a). Some utilities hardly have an information management system in place to meet this requirement and installing one requires large investments at a time when some utilities are struggling to finance capital works with the current water tariffs.

Section 121 is even more fiercely debated. This section is about punishment for contaminating water courses. To quote, this section reads 'a person who contaminates or causes to be contaminated any watercourse of the water supply system or any part of the watercourse or water supply system with any substance' shall be liable to imprisonment for a term not exceeding 20 years or the death

---

[21] Personal interview on 8 May 2009.

[22] Personal interview on 8 May 2009.

penalty if death is the result of his or her action (MEWC, 2006a: 92). Water utilities questioned this section (and its harsh punishment) as they find it unjustifiable if such occurrences were to take place outside their control. Moreover, as with this section, water utilities claimed that they were powerless in this respect as responsibility over water resource protection lies with state governments.

### 3.5.2 Water resource management

Changes in water resource management were excluded from the reform. Thus, the reform only dealt with matters related to the management of the water supply services (as stipulated) under the Concurrent List. However, as the entire value chain for water also encompasses water resources, any deterioration in (raw) water quality will have an effect on water supplies. Here the reform demands that state governments play a pro-active role in protecting water resources, through their state water resources regulator. As a state department, the state water resource regulators' effectiveness in carrying out its duties reflects the political priority that the state governments give to preserving water resources.

Evidence from the interviews revealed a general feeling that state governments and water resource regulators were not adequately protecting water resources. Some water utilities and civil society groups fiercely criticized the state water resources regulators for failing to enforce environmental laws. Representatives of several water utilities[23] and non-governmental organizations[24] felt that state governments lack the political will to control sand mining in rivers and to gazette (protect) water catchment areas. A chief engineer with JBA Perlis did not 'see why the state government is not taking up this issue seriously'[25] despite having full control over water resources and commercially benefitting from (raw water) royalties. He added that if the state government's lackadaisical attitude towards protecting water resources continues, more and more rivers will be polluted, eventually reducing the availability of (raw) water for treatment. In the past, severe drought has affected water supply in several states, such as Melaka, forcing them to import raw water from neighboring states. In the 1990s Melaka had to import water from Muar River in Johor to fill its Durian Tunggal Dam (Angkasa Consulting Services, 2011). At present the biggest ever Malaysian water transfer project[26] costing more than RM 10 billion is being implemented, pumping 1,890 million litres per day of raw water from rivers in the state of Pahang to meet future demand of the most developed (but water-stressed) state of Selangor, the Federal Territory of Kuala Lumpur and Putrajaya until 2025 (MEWC, 2010).

Several organizations, particularly FOMCA[27] and CAP[28], have repeatedly demanded that water resources be included within the spectrum of the water reform. Both believe that only the federal government has the political means to effectively safeguard water resources and should take over

---

[23] Personal interview with several water utilities.

[24] Personal interview with several non-governmental organizations.

[25] Personal interview on 28 Apr. 2009.

[26] Expected to be completed in 2014.

[27] Personal interview on 10 Aug. 2009.

[28] Personal interview on 1 Sept. 2009.

the management of water resources from state governments. Within the legal system in Malaysia, this can only be done by amending the Federal Constitution. At the time of writing, their requests have not received much support. An Executive Director of the NWSC[29] points out that amending the Federal Constitution would require the support of state governments and he found it very hard to believe that state governments would voluntarily surrender their sovereignty over water resources. Equally, the federal government wants to maintain cordial federal-state relationships and will not push very hard for such an amendment. Even if a new political landscape emerges after the 13th General Election[30], it is unlikely that changes in the existing structure of water resource management will happen. Perhaps a more feasible route at the present is to ensure that state water resource regulators get the support they need, both politically and financially. Politically, they must be allowed to act freely and prosecute polluters. Financially, state governments, such as Kedah[31], must be fairly compensated for losing business opportunities if they gazette water catchment areas.

### 3.5.3 Financing

The water supply industry is a highly capital intensive business venture. Venturing into business requires enormous initial investment for building treatment plants and reservoirs, and laying down a transmission network with pipes, pumping station, etc. (Prasad, 2007b). Acquiring investments of such magnitude drives most governments (in under-developed and developing countries) to turn to international or private donors. In most cases, private participation (mostly by foreign companies) is included as one of the conditions in exchange for loans. However, treating water as a commodity only benefits those who can afford to pay and can exclude the poorer sections of the community (with less ability to pay) from the water supply (Kessides, 2004). At the same time public water utilities are struggling to extend their network coverage and to maintain water infrastructure.

The reform established PAAB to address the financial constraints facing the water sector. It is estimated that the water sector will require RM 110 billion for water infrastructure development and maintenance over the coming fifty year period (MEWC, 2008). This is a huge amount of investment and raising this much will certainly be a challenge to PAAB. With backing from the government, many believe that PAAB has relatively easy access to the financial market, but many water utilities worry about the foreign exchange risks. According to the Director of JBA Kedah[32] this can increase the cost of borrowing and will lead to water utilities being charged higher lease rentals. To minimize this risk, CAWP[33] has proposed that PAAB should source funds from the local financial market. Notwithstanding this discussion PAAB's financial approach has been proven to be an important ingredient within the reform. A representative of the NWSC anticipated that it was quite difficult to predict whether 'in the future, water utilities can undertake and finance water projects independently and without resorting to PAAB's financial assistance'[34] This can

---

[29] Personal interview on 9 Sept. 2009.

[30] Expected in 2012/2013.

[31] Expressed by a state politician in an interview on 19 July 2010.

[32] Personal interview on 11 May 2009.

[33] As contained in its memorandum to the Minister dated 27 Feb. 2006.

[34] Personal interview on 23 Sept. 2009.

only happen when water utilities become sufficiently financially independent (to service loans repayment), something that might only be realized within a 30-year timeframe (MEWC, 2008).

Another contentious issue is the effect of the financial mechanism on water utilities. Some water utilities were still unsure about how this would pan out. For instance, the water utility SATU[35] did not know if PAAB would eventually take over its operations. By the same token, a senior engineer with the water utility LAP[36] questioned the merit of subjecting all water utilities to PAAB without taking into consideration the financial ability of the utilities to service their lease payments. These initial doubts faded away as soon as PAAB demonstrated its willingness to accommodate different levels of readiness (of paying the lease payment) among water utilities. PAAB has developed a flexible approach, categorizing water utilities into four categories, depending on their financial standing and the different levels of lease payment they have to pay (PAAB, 2007). Category One consists of water utilities recording an operational loss. Category Two includes water utilities recording operational revenue that is insufficient to pay the full lease rental. Water utilities in these two categories, mostly public water departments, are recognized to be unable to pay the full lease payment and will receive government subsidies to cover the shortfall. Water utilities in Category Three are those (mostly corporatized water entities) with the capacity to pay the full lease. Lastly, Category Four represents water utilities operating on concession contracts (in Selangor and Johor). These are financially independent utilities assumed to be able to pay lease payments on (full) commercial terms. It is argued that PAAB must uphold the principle of affordability in determining the lease payments. This means that the lease payment charged will have to stay below commercial rates, especially for water utilities in Categories One and Two. It is imperative that the federal government continues to fund PAAB, until the water sector becomes efficient and effective.

### 3.5.4 Operational issues

Since 2006, the reform has accelerated the corporatization process: four state water departments have been corporatized and another one is to do so in 2012. Including the utilities that were corporatized prior to the reform, only two water utilities – JBA Labuan and LAP – now remain to be corporatized. Both these utilities are still awaiting final consent from their state governments. An important feature among these corporatized water entities is that they remain under the sole control of the state government, even though they are commercially managed. This means that the business of water supply is (and perhaps will be kept) under public control. This, to some extent, has helped to allay fears among several NGOs, such as FOMCA, CAP and CAWP, about turning water over to private hands. However, this structure has affected the opportunities available to some water utilities who wanted to participate in the entire value chain of the sector (treatment, distribution, collection). Water corporations (and not private water utilities) will become the dominant structure in the water sector.

Some private companies consider corporatization as a 'threat' to their existence. During one interview, a representative from a private water company expressed dissatisfaction that the reform

---

[35] Personal interview on 6 July 2009.
[36] Personal interview on 22 May 2009.

has dented their ambition to expand business through (full scale) privatization.[37] Some even regarded this as an attempt to nationalize private water assets. Although the respondents were not strongly opposed to corporatization *per se*, they urged the government to respect the contracts that their companies have signed with state governments.[38] The federal agency has taken an open approach in assessing the contribution that the private sector makes (and should make) in the water supply sector. The Economic Planning Unit (of the Prime Minister's Department) and the NWSC, for instance, suggested that greater private participation could be considered when the water sector has been stabilized. This is in line with the government's policy of facilitating private sector involvement in the country's economic development and growth (Economic Planning Unit, 2006).

Despite these controversies it can be concluded that the institutional reform (corporatization) is progressing well and is contributing to the objective of the reform to make the water sector more sustainable in the long run. It is important to keep a close eye on how these newly corporatized water utilities reinvent themselves in further supporting the objectives of the reform.

## 3.6 Conclusions

The historical trail of the water sector reform started as early as 2004. Several events and driving forces accelerated the reform towards the results we see today, with several new institutions and laws. The reform process stemmed from the recognition – by federal and state governments alike – that a 'business as usual' approach to water provisioning was no longer tenable. The radical transformation program (the reform) will bring the water sector more in line with other utility sectors such as telecommunications, energy and gas in terms of investment, regulatory oversight, service standards and public perception.

Institutional reform has altered the landscape of the Malaysian water sector. As analyzed in Sections 3.4 and 3.5, the reform marks a significant move towards establishing good governance in the sector, especially in having a clear division of tasks between policy formulation, regulation and service provision. These tasks have now been assigned to a wide spectrum of both state and non-state actors, each having a clear role, responsibility and job description (Table 3.3). Nevertheless, the analysis of the efficacy of the output produced mixed results. Overall, the output was crucial for the reform, but their contribution to reaching some of the objectives has been contested. Many stakeholders believe that freedom from political interference and access to (reliable and quality) information (see Chapter 5 and 6) is essential for the regulator to perform effectively. Equally, the water resources regulator (established to complement the role of the regulator on water resources) can only function effectively with adequate political and financial support. Corporatization has progressed considerably through the reform, although higher efficiency levels will only be injected into water corporations where regulatory oversight has matured and there is a viable financial mechanism in place. In this respect, there are still uncertainties about (foreign) borrowing risks, the mechanism for determining the lease rental and the timeframe in which full cost recovery can be achieved.

---

[37] Personal interview on 6 May 2009.

[38] Personal interview on 13 May 2009.

*Table 3.3. Governance institutions in the water supply sector (MEWC, 2008).*

| Body | Area of responsibility | Description |
|---|---|---|
| Federal government | Overall policy | Develop a holistic water policy for the country |
| State governments | Water basins | Manage existing water basins with the view of protecting the quality of raw water and identifying new water basins when required |
| NWRC | Governance matters | Ensure co-ordination with state governments in the management of water basins |
| NWSC | Regulation | Regulate the whole water and sewerage industry based on policy directions set out by the federal government |
| PAAB | Management of water assets | Acquire finance and build water assets based on 'build and lease' model |
| Water operators | Water supply matters | Operate and maintain water assets based on 'asset light'[1] principle |

[1] Refers to a situation where water utilities are no longer responsible for the development of water assets, but instead lease them from PAAB. Hence, water utilities become 'asset light' companies, focusing purely on the (core) business of (delivering) water supply.

# Chapter 4.
# Water supply reform: a policy arrangement analysis

## 4.1 Introduction

In this chapter, the water supply sector reform is analyzed from a policy arrangement perspective. The analysis aims to increase understanding of the reform process through an in-depth analysis of the policy processes and public interventions that occurred during the reform. The analysis is based on the dimensions of the policy arrangement approach (see Chapter 2). Section 4.2 discusses the discourses that dominated the reform process. Section 4.3 analyzes actor-resource relations. In Section 4.4, two set of rules (of the game) are analyzed: the (pre-existing) rules regulating the reform and the rules that emerged as a result of the reform. Together these sections provide an in-depth understanding of how the water supply reform process took place, which is summarized in the conclusions in Section 4.5.

## 4.2 Changing water supply discourses

Generally discourses in the water supply sector – in Malaysia, as well as other countries – are formed around two axes. The first axis is that of state versus privately run water supply systems. Along this axis three dominant discourses could be distinguished during the water supply reform process in Malaysia; briefly identified as public, private and corporatization. The second axis concerns the decentralization and centralization of authority in the water supply sector. This axis had two main discourses in Malaysia. The first focused on the devolution of authority towards the level of the states and was sometimes also closely linked to the corporatization of state water companies. In the second discourse power and authority in the water supply sector is preferred to be centrally located, at the federal state level.

### 4.2.1 First axis: state versus privately run water supply systems

Since independence, two main discourses have dominated the water supply sector in Malaysia. The first discourse largely concentrates on the central role of the state in water provisioning. The second concerns the growing importance of the private sector in water provisioning, often in response to the inefficiencies of a publicly-run system. With the country's progress and economic growth, and (the perceived) failures of both systems to satisfy the demand for efficient and sustainable water services (as also discussed in Chapters 5 and 6), a third discourse emerged, concerning the corporatization of water supply. This discourse can be interpreted as a compromise model between the two other discourses. The three discourses will be introduced below.

*Public water supply system*

In almost all countries the water supply system is always developed by a public water services company or a local government. In the UK and France, for example, water supply provision was for a long time a function of the municipality or of local authorities (Rouse, 2007). The public water supply system still dominates in Africa, Asia and Latin America. In the discourse on public water provisioning, sufficient, safe and affordable water is seen as a basic good, which the entire population should be entitled to. Because of this water supply should be kept out of the hands of market parties, with their search for (short term) profits. This discourse views the public water supply system as being capable of fulfilling the social and public health policy goals of the state: providing safe water at reasonable costs for the entire population. In the twentieth century, the dominant discourse of the public water system entailed a strong social contract between the state and its population (Mustafa & Reeder, 2009). The social policy and social contract characteristics of public water supply systems have prevailed for quite some time in the Malaysian water sector.

Immediately after Malaysia's independence in 1975, the provision of water supply was assigned to the Public Works Department (PWD), a federal department under the Ministry of Works. At that time, water supply was implemented to meet the social policy of providing the population's basic need for water and promoting public health. However, as time went by, the provision of water supply became transferred to the state level. This transfer marked an important milestone in public water supply, with the establishment of state water departments. The states attempted to realize a social contract with their populace by providing a consistently good quality and affordable water supply. There were two main criticisms of this model. The first is related to public financial resources that needed to be spent (even though they are often not adequately available) to ensure that the system fulfilled its promises of providing affordable, high quality water for the entire population. In dealing with this challenge some Malaysian state governments delegated the service to private companies with sufficient financial resources, while others continued to rely on limited financial resources from the federal government to run water supply services. This often resulted in them not meeting set objectives. Neither strategy was able to satisfactorily improve the efficiency or quality of service of the Malaysian water sector (see Chapters 6 and 7). The second discussion involves the question of separating water regulating and policy-making from water services and supply. Some argue the need to keep both tasks in the hand of the state, while others see the benefits, possibilities or even necessities of separating these tasks between public and (semi)private bodies.

*Private water supply system*

The discourse on water privatization has been a global phenomenon, and to a major extent was a hostile reaction to the shortcomings of public water provisioning. After the successful (but controversial) privatization of many public services by Margaret Thatcher's administration in the UK in the 1980s, the discourse and practices of privatization of public services, including water supply services, spread to other parts of the world (Rouse, 2007; Prasad, 2007a). In the late 1980s, developing countries in Africa, Asia and Latin America started to implement privatization. This process accelerated tremendously during the 1990s and 2000s when it was taken up in newly emerging economies, such as China, Brazil and Argentina (Prasad, 2007a). After the collapse of the communist regimes in Eastern Europe and the USSR in 1989-91, privatization has also been making inroads into Central and Eastern European countries. Since the 1990s onwards, large parts of the global private water sector have fallen under the control of just three or four French corporations, their subsidiaries or partners (Robbins, 2003).

Despite geographical differences the main motivation for privatization has been tied to the grounded belief that the private sector is more efficient and cost-effective, and that privatization provides a favorable economic climate for a water enterprise to excel through increased competition. Usually privatization is accompanied by the institutionalization of a regulatory authority, aimed at regulating the behavior and performance of the private water companies. In the UK, for instance, the UK government established three regulatory agencies undertaking different roles: the OFWAT as the economic regulator; the Drinking Water Inspectorate (DWI) for drinking water control; and the Environment Agency (EA) for environmental regulation (Dore, Kushner & Zumer, 2004). In other places, privatization has usually been accompanied by a separation of regulatory tasks and provisioning tasks.

The discussions about privatization in the Malaysia water supply sector were part of this global discourse over the participation of the private sector in utilities. However, the privatization mode implemented in Malaysia has differed from that of the UK in terms of asset ownership. Under the UK model, privatization involved disposing of public water assets to the private sector (Silvestre, 2012). Under the Malaysia model, the private sector was only given the right of use of (public) water assets, while the state remained the owner of those assets (Hukka & Katko, 2003). While the UK model resembles full or material privatization, the Malaysia model is more of a public-private partnership, of which there are several possible forms such as BOT, BOOT and concessions. Under such arrangements, the involvement of foreign water companies in the domestic water sector has varied: from technical service agreements (e.g. between Thames Water and SAJH), BOT (e.g. Suez in Sabah, Veolia in Perak and Selangor) and concessions (e.g. Thames Water in Kelantan) (Hall, Corral, Lobina & Motte, 2004).

Following this discourse, Malaysian water companies have also become involved in the privatization of water provisioning, in both local and international water markets. For instance, SAJH, SWC and ESB dominate the Johor water sector, while PNSB, Splash, ABASS and Syabas are main water providers in Selangor. Taliworks runs the water system in Kedah, while AUIB and Salcon are involved in BOT concessions in Kedah and Negeri Sembilan, respectively. Some of these Malaysian companies are also venturing into international water markets. Salcon, for instance, runs a water treatment plant in Changle, Shandong Province, China, and has a bulk water supply contract with the Linyi Municipality in Shandong. YTL manages Wessex Water in the UK (Hall *et al.*, 2004) and Ranhill Utilities, a parent company of SAJH, manages several water treatment plants in Thailand and China, and has started to made an inroad into Saudi Arabia's water sector (Ranhil Utilities Berhad, 2007).

The failures of the Buenos Aires and the Cochabamba concessions – the world's largest concession and the cause of the world's first water war respectively – have invited several fierce criticisms of water privatization (Casarin *et al.*, 2007; Ahlers, 2010). In some parts of the world privatization has failed to achieve two fundamental objectives: the fiscal objective – the inability of privatization to relieve governments of the burden on investment financing, and the efficiency objective – that water utilities performance have not improved under private ownership (Araral, 2009). In addition, in some places private water companies are only interested in serving the urban rich, and neglect the rural (and urban) poor, and have been accused of charging higher prices than publicly managed utilities (Dore *et al.*, 2004). They have been blamed for showing little interest in expanding their network to areas which they consider to be unprofitable, such as slums and rural areas, but demonstrating great interest in improving bill collection (Robbins, 2003). These critical assessments of privatization in the water supply sector have also been voiced in Malaysia. Abbott, Wang and Cohen (2011) claim that while privatization has managed to increase supply coverage to almost 100% in Selangor and Johor, consumers have not substantially benefitted in terms of lower prices. Water tariffs in these two states are the highest in the country (MWA, 2011). The water privatization in Kelantan was considered a complete failure. Thames Water was forced to exit the Kelantan water sector after recording a massive RM 100 million loss. The Kelantan state government was forced to step in and subsequently bought back the concession from Thames Water (Hall *et al.*, 2004). Generally, failures of privatization have been due to an absence of a viable financial model (accepted by water companies and water users) and any effective regulation of the private water supply sector (MEWC, 2008). The privatization of water supply is now subject to the same level of critical scrutiny that was paid to the public water supply sector a decade ago.

The failures of privatization (and of the public water system) prompted Malaysia (and many other countries) to examine alternatives for developing the water supply sector. One of the alternatives considered was to embark on an institutional reform of the public water sector; a process in which the private sector culture is introduced into a public water sector. In general terms, this process is referred to as corporatization or commercialization (of the public water department). This leads us to the third discourse: corporatization.

## Corporatization

As a reaction to the debates on and experiences with state hegemony of the water sector and neoliberal water supply models, a third alternative discourse on corporatization developed. This discourse emphasizes reforming the institutional set-up of a public water supply system by bringing in institutional arrangements and management practices associated with the private sector in the institutional context in which public water companies operate (Schwartz, 2008). While advocating keeping the water supply within the public domain, (which distinguishes it from the privatization model), the corporatization discourse embraces the strengths of corporate models within public water supply organizations. In a nutshell, this model seeks to combine the strengths of the public – in public policy – and the private sectors – in working culture and performance. While maintaining control over water resources, state governments gradually bring in outside expertise and knowledge, and adopt private sector working culture, such as output and outcome-based targets and establishing a link between pay and performance (Siddiquee, 2010). Others refer this model as new public management, a new approach to improving public sector service delivery (Brown, Ryan & Parker, 2000; Schwartz, 2008).

The reform of the Malaysian water sector highlights the importance of the corporatization discourse, which has helped to re-define the water supply sector. It has re-iterated the fundamental role of water as a public good, while at the same time recognizing the economic value of water.

This discourse gained an overwhelming acceptance from Malaysian state governments[39], an indication that it fits well with the current conditions and preferences in Malaysia. Moreover, corporatization has reduced fears among civil society organizations (such as CAP, FOMCA and PCPA), as it guarantees that water will remain in public hands and it limits private control over this precious resource. This re-affirmation of public control over water does not, however, close the door on public-private partnerships and collaboration (e.g. service contract, management contract) in the water sector, as these do not involve transfer of ownership.

The strong position of the corporatization discourse discredited the authority and dominance of the private water supply system in contemporary reforms of the Malaysian water sector. For example, an Executive Director of a private water company in Kedah indicated that the reform has reduced the likelihood of them participating in the (full) privatization of the water supply sector in the state.[40]

A key element of corporatization is the separation between policy/regulation and service provision. This discourse holds that a natural monopoly like water is best regulated by the state (Prasad, 2007a), with the service provisioning allocated to financially independent corporate-like bodies. This separation of functions reduces the chance of conflicts of interests arising as these two functions are assigned to two separate bodies – the central water regulator (in Malaysia the NWSC) and corporatized water provisioning bodies.

Centralizing regulation within a federal state authority has caused states to re-define the role of their lower level regulating bodies. Three clear examples illustrate the different ways in which the role of the state regulatory body changed. First, the (existing) state regulator has been forced to give up its regulatory tasks (as in the case of BAKAS). Second, the state regulator was forced to take on new regulatory tasks (as in the case of BAKAJ). Third and lastly, a new state water resources regulatory body is formed (as in the case of LUAN). Within these changes of the state regulatory body is especially important to understand how effective water supply regulation can be achieved. Even though the hegemony of the central or federal regulator in water regulation is obvious, the enforcement of such regulations depends at least on two realities. The first one is that water regulation and thus water supply is becoming increasingly political and thus influenced by politics (Marques, 2006). For instance, interviews with respondents revealed that many of them had arrived at the common conclusion that the more the central regulator depends on political powers, the higher the likelihood of politicians exerting control and influence over the central water regulator. Many interviewees cited the appointment of a political board of members of the regulator as an example of this. Although, at the time of writing, there is still limited political influence on the central water regulator, no one can guarantee that such interference will not enhance in the future. The second reality is the information asymmetry between the regulator and the regulated water companies. Most of the information in the water sector is in the possession of water companies. Realizing the importance of having reliable information, the NWSC devoted a substantial amount of time to data gathering when it was established.

---

[39] Five state water departments – Melaka, N. Sembilan, Kedah, Pahang and Perlis. Others were corporatized prior to the reform.

[40] Personal interview on 6 May 2009.

### 4.2.2 Second axis: decentralization and centralization of the water supply sector

Two main discourses can be clearly distinguished along this axis in respect to the Malaysian water supply sector. The first – the decentralization discourse – focuses on the devolution of authority towards the states or even the corporatized water companies. The second discourse, of centralization, emphasizes the importance of central/federal power and authority in the water supply sector. The decentralization discourse emerged strongly in the policy arrangements during the pre-reform era – before 2006 – and is traditionally closely linked to the dominant role of the states in the sector. The centralization discourse became an essential part of the reforms when the government tried to improve the efficiency of the water supply sector. Let us investigate both discourses in more detail.

For more than five decades, the provisioning of water supply has been one of the key public policies of the state government. As enshrined in the Constitution, the provision of water supply has been the sole responsibility of the state. As a consequence, every state established a water department or water corporation to manage almost every aspect of water supply – from abstraction to bill collection. Even though some activities were recently outsourced to the private sector, decentralization of the water supply to state authorities was a central feature of the system. Under the decentralization regime, each state plays a key role, and as time went by they have developed their own forms of response to socio-political needs (Zaini, Rakmi & Aznah, 2008). Any attempt to challenge this state dominance was perceived as an attempt to undermine the sanctity of the federal-state relation, guaranteed under the constitution. This was the reason why the state governments rejected the earlier idea of centralizing water management (in 2003), especially as it failed to acknowledge the states' power over water. This model, however, has been criticized as contributing to inefficiency and hindering the separation between policy/regulation and service provision. In other words, the dual role of state governments – being in the water business and at the same time being water regulator – has the tendency to allow conflicts of interests to occur. From the perspective of (good) governance, state governments that are involved in the water business have less motivation to undertake independent and effective regulation. Because of this the notion of effective regulation became an essential part of the reform process. The aim was to achieve this by taking regulatory power away from the states and placing it under a central body at the federal level. This is the main thrust of the centralization discourse of the water supply, as discussed below.

The centralization discourse emphasizes the benefits of central management and regulation of the water supply sector, and particularly the role of the federal government. Such centralized management is intended to facilitate effective regulation and a sustainable financial model in the sector – two fundamental problems of the decentralized model – while re-affirming the states' control over water. It is based on the arguments that (a) effective regulation is needed to curb the potential of market abuse by market actors in a monopolistic water industry, and (b) that as a public good water must be made available and affordable to every consumer through a viable financial model. Thirdly and finally, effective regulation and a viable financial model can be best managed by a central government body.

The centralization discourse has re-defined the general understanding of central management in water supply. The previous understanding equated centralization with the federal government having total control of the water sector; a proposal that was rejected by the state governments in 2003. The new understanding of federal involvement in water supply adopts the principle of 'shared-responsibility'; recognizing the roles of both the federal and state governments within water management. The first part of the equation – water regulation and financing – is put under the jurisdiction of the federal government, while the second part – water resources and service provision – remains under the control of the state governments. This has the benefit of ensuring the transparent separation of policy/regulation from service provision, which has been further developed in practice by the establishment of two central bodies – a regulator (the NWSC) and a financier (PAAB) – to oversee these responsibilities, while state water departments were converted into water business outfits (through corporatization) to manage water supply.

In summary, the corporatization discourse has strengthened the dominant role of state governments in water provisioning, albeit in a more modern way where a business culture has been introduced and the tasks of regulation and provisioning have been separated. This same discourse also undermined the calls for water privatization and diminished the role of the private sector in water supply. Along the second axis, the centralization discourse became more dominant, but took a new form, with recognition of the need for shared responsibility between the federal and state governments. It has also allowed for federal intervention over regulatory and financial mechanisms, delegitimizing calls for a purely decentralized organization of water regulation and supply.

## 4.3 Actors and resources in the water reform struggle

The reform was dominated by the interaction of three groups of actors: state actors, private actors and civil society actors. While the definitions of private and civil society actors are self-evident, state actors can be further categorized into two broad categories: the federal government (and its public bodies and organizations) and the state governments (and their bodies and organizations). The degree to which each category of actors could influence the outcomes of the reform was dependent upon the types of powers and resources they possessed and were able to use in the reform struggles and their interactions with other actors.

### 4.3.1 Federal government actors

Generally we can understand the reform as an attempt from the federal government to improve the functioning of the water supply sector, in line with its constitutional obligations. The water supply reform was a state intervention, in which the federal state actors, specifically the MEWC, Economic Planning Unit (EPU), Ministry of Finance (MoF), Attorney General's Chamber (AGC) and Department of Environment (DOE), were the prime initiators. Despite each of these federal agencies operating and being guided by their own set of rules, their actions were, to a certain extent, also inter-related. The different federal agencies depended on each other in order to function effectively. This suggests a need to jointly mobilize or share resources in order to mutually strengthen their power.

The reform was spearheaded by the MEWC, a body dedicated to coordinating water management in the country. To initiate the reform process, MEWC decided to involve the services of an external expert, especially for developing the business model for the water sector. The services of an outsider were sought for two reasons: first, MEWC did not possess the required knowledge about the economic aspects of water supply; and secondly because there were major time constraints and additional input was needed. This resulted in the appointment of a group of consultants to conduct a preliminary feasibility study on the proposed reform. However, as already explained, no single actor could act independently. The appointment of the consultants involved a power play between many of the actors. First, the MEWC needed consent from the MoF on who could be appointed to undertake such a study. Second, the EPU had to approve allocation of a budget for the study. Without support from both these agencies, it would have been difficult for the MEWC to initiate follow-up actions, (which subsequently resulted in the formulation of the WSIA and SPANA: the main legal outputs of the reform). This clearly shows that resource-dependency does not need to be a constraining factor in policy arrangements.

When no single actor has absolute control over the available resources, it is unusual for one actor to determine the policy process. In such a situation, actors are better positioned to influence the outcomes of the decision making if they collectively bring their resources together (Arts & Van Tatenhove, 2004). The process of establishing the WSIA and SPANA clearly indicates how the resources of various actors were pulled together and mobilized to attain a policy objective. For instance, the MEWC needed to tap into external resources – financial, legal, information, expertise – in order to realize the reform process. Hence, the MEWC was forced to cooperate with other federal actors and share resources. Perhaps the most crucial issue concerned Cabinet decision-making, required to approve and legitimize the reform. The MEWC also subsequently, collaborated closely with EPU on the development of water policy, with the MoF for fiscal and monetary policy, with the AGC on legal inputs and with the DOE on the environmental dimensions of the policy reform. Discussions among these federal bodies, with the state governments (on state water enactment and water resources extraction) and the water utilities (on operations) greatly expanded the knowledge horizon. This was further extended by studying various reform models in other countries. The Minister of the MEWC led water sector study visits to the UK and France. The UK model, especially with respect to the regulation oversight, was thought to be the most appropriate for Malaysia. The establishment of the NWSC was strongly modeled upon OFWAT, the economic regulator for the water sector in the UK and Wales.

However, the need to share resources and power can also create unstable situations, as a result of power struggles and confrontations between different state sectors. Tensions arose when the shared objective (of the reform) collided with individual departmental objectives. A study of the reform process illustrates at least two examples of this. First, the way the reform sought to 'outlaw' further water privatization (and instead favoured corporatization), contradicting the general pro-privatization policy adopted by the EPU. Second, the reform was under constant pressure for not adequately handling environmental issues, as expected and advocated by the DOE. The Deputy Director-General of DOE stated that the reform 'skewed towards the water side and gave less focus to the sewerage side'.[41] These confrontations demonstrate a certain conflict of interests among federal actors, which ran the risk of delaying the reform process. Prolonged confrontation can harm actor coalitions if it is not resolved, especially when such confrontation involves crucial policy issues.

In this respect, the MEWC knew that it did not have the relational power (power to influence others) to force a harmonious co-existence between these actors in the course of the reform. Thus,

---

[41] Personal interview on 16 Oct. 2009.

an intervention from a third party with a greater authority was needed to mitigate confrontations. Here the MEWC used its relations with the country's top decision makers – the Prime Minister and his Deputy – to forge harmonious relations among federal actors in the reform process. Remarkably, the MEWC was able to get both the Prime Minister and his Deputy to preside over three high level meetings throughout the reform process. All these meetings, which took place in 2006 and were initiated by MEWC, were attended by all the top officers from the key federal agencies. The attendance of the Prime Minister and his Deputy at these meetings reflected their commitment and the desire of the government to handle this matter in an amicable and smooth way. The authoritative power of the Prime Minister made it possible to resolve many policy issues, such as the corporatization of state water departments, the establishment of the financial model, the centralization of regulatory powers and the state retaining power over water resources. The meetings subsequently paved the way for the amendment of the Federal Constitution, which eventually formed the essential part of the legal framework supporting the reform (the WSIA). The formulation of the WSIA reflects the strong influence that the MEWC had over the outcomes of the reform. The combination of the dominant power of the MEWC and the influence it had on the top decision makers during the three meetings expedited the successful formulation of both the WSIA and SPANA by the federal state and their successful passing by Parliament in 2006.

### 4.3.2 State government actors

The role of the state government actors in the reform process can be discussed from two different perspectives: vertical relations and horizontal relations. The first perspective concerns the relations between state actors and federal actors; the second concerns the relations between the different state actors. In a broad sense, these state actors can be grouped into three categories: state governments; state water departments or corporations; and state water resources regulators.

*State-federal state relations*

The Federal Constitution divides the legislative jurisdiction into three lists: the State List, for matters exclusively under the jurisdiction of the state government; the Federal List, for matters exclusively under the jurisdiction of the federal government; and the Concurrent List for matters where both Federal and state governments have joint powers to legislate and decide (MEWC, 2008). Prior to the 2005 amendment to the Constitution, the jurisdiction and control over water (including water resources, supplies, rivers and canals) were under the State List. Each state had its own separate legislation relating to water, only applicable in that particular state.

The 2005 amendment to the Constitution facilitated the shared-responsibility principle of the reform: 'water services' were taken off the State List and placed on the Concurrent List, thus allowing the federal government to have an active role in the water sector, especially with respect to regulation. The growing role of the federal government, however, did not diminish the power of the state government over water resources, nor did it change the role of the state governments as the sole owner of the water company after the corporatization of the state water department. In other words, the reform recognized the importance of upholding state control over water resources and the necessity of having effective regulation by the federal state. It was believed that the reform would not function well or would be incomplete if one of the two powers (state and federal) was missing.

Despite having legal responsibility over water resources, state governments did not have the financial resources required to effectively safeguard water resources (in terms of quality and quantity), nor to fully meet their obligation to provide water to the entire population. Thus, their power over water resources was undermined by their dependence on federal actors for financial resources. In many areas, such as financing and regulation, federal actors remained the dominant forces (and even more so after the reform). Federal systems (such as Malaysia) often have to grapple with the division of powers and responsibilities between the federal regime and state authorities. In Malaysia federal agencies (such as the EPU, the MoF, the AGC, the DOE and the MEWC) play a dominant role in their respective fields (economic development policy, fiscal and monetary policy, legal expertise, environmental policy and water infrastructure project implementation). These federal actors used their control over resources as a 'weapon' to subjugate the state governmental actors to accept the reform. For instance, state agencies accepted the corporatization (of state water departments) because it would allow them access to financial assistance from the federal state.

The state governments were not just weak in terms of their influence on the outcome of the reform; they also lost their powers of regulation after 2007. The move to centralize the regulating power for water to the federal level in 2007 was the result of weak and ineffective state regulation over water, a situation that had been caused by the conflicting roles of state governments as owners, operators and regulators in the water sector (Casarin *et al.*, 2007). The state regulators were not in a strong position to effectively enforce the mechanisms needed to improve efficiency in the water sector, and this severely jeopardized any attempts to improve overall service quality. Centralizing water regulation has, to a certain extent, led the state regulators to take on a new, less prominent, role. Under the new rules of the game, the federal government dominates the regulatory regime and state governments have no power to challenge the legitimacy of these decisions. In most cases, the regulating role of state governments has been reduced to safeguarding water resources.

## State-state relations

This section describes the relations between state governments, the state water companies and the state water resource regulators.

As explained above, the corporatization of the state water departments did not take away the role of the states over water supply. What it did was to transfer the regulating power of the state regulators to the central regulator; a shift triggered by weak and ineffective state water regulation. By the same token, the state water corporations remained state bodies, obliged to adopt a commercial and business culture and financial independence, following the private sector model. This marked a significant shift from typical public water operations. The shift towards greater efficiency needed to be accompanied by a certain level of autonomy and a separation of powers. Corporatization implied the states relinquishing their traditional control over the water department. The water corporations were granted autonomy over the management of financial matters and human resources. However, in contrast to full privatization, the state governments remained the sole owners of the company. The dual roles of the state governments (as regulator and as service provider) were broken by the establishment of corporatized water services. The separation of the two functions became clearer, with the state governments (through their respective water companies) assuming the role of a 'service provider', while responsibility for overall policy and for regulation of the sector lay under the jurisdiction of the federal government, with the exception of the regulation of water resources.

The state water resource regulators are the last remaining manifestation of state power over water resources. They are usually fully-fledged state bodies that assist the state administration in protecting its water resources. Their powers are derived from the local state, which provides its resources (human and financial) from the state government. However, their ability to safeguard water resources from uncontrolled economic activities, such as logging or sand mining, is the subject of severe criticism from civil society organizations. The Malaysian Nature Society[42] has been critical of state governments' ineffectiveness in safeguarding water resources from such activities and their reluctance to gazette water catchment areas. This ineffectiveness has been attributed to two main reasons. First, as highlighted by the Chief Operating Officer of FOMCA[43], state governments lack the political willingness to stringently enforce environmental laws as this can erode their revenue from economic activities, such as logging. In this respect it seems that state governments are trapped between economic gains and environmental protection. However, one state politician[44] in Kedah revealed the readiness of his state administration to protect a catchment area, provided that the federal government compensated them for loss of income. He added that the federal government had not responded to this request. Second, water resource management at the state level involves several other actors, since water is also needed for other sectors – irrigation, mining and fisheries. In fact irrigation (for the cultivation of rice) uses more than 60% of the country's water resources (Economic Planning Unit, 2000). State water resource regulators lack the power to control other actors that make claims on water supplies and have to rely on the resources, especially legal ones, of other federal departments (Agriculture, Drainage and Irrigation, and Fisheries) to address such issues.

### 4.3.3 Private actors

Private water operators are the most important private actors in the water supply sector. Before the reform was introduced, there were close relations between them and the states that extended to almost every aspect of water supply. This relationship was centered around the participation of the private sector in the states' water sector and the legal control of the states over water. From raw water abstraction permits to treatment and distribution licenses, the private water operators were dependent on the state and over time they came to be involved in many activities that were previously the function of the public actors. Their participation in the sector ranged from small-scale short-term service contracts to large-scale long-term BOT, BOOT and/or concession contracts, involving treatment, distribution and revenue collection activities. The water concessions in Selangor and Johor demonstrated the dominant role of private water operators. In both states, they had been involved in all aspects of the operations: from treatment to revenue collection,

---

[42] Personal interview on 11 Aug. 2009.

[43] Personal interview on 10 Aug. 2009.

[44] Personal interview on 19 July 2010.

for as long as 30 years. Nevertheless, the state kept a firm grip on the private sector through regulation. Despite the private companies being financially rich, the states had discretionary power over them, particularly as they had the power to decide whether (or not) to renew their contracts when they expired.

The reform has affected two aspects of the role of the private water operators: in finances and preferential positions. First, it has diluted the dominant power of the financially-rich private water operators. Their position was challenged with the establishment of the PAAB (the water asset management company), a government financier for the sector. Costly private financing was (partly) blamed for high water tariffs, as private water operators rushed to re-coup their investment and make a profit on these via high water (treatment) prices. One of the major goals (and achievements) was to establish a viable financial model which could provide substantial public funding and mitigate excessive tariff increases from private financing. The PAAB was established to provide financial resources to the water sector. Equipped with huge financial resources, PAAB has revolutionized the way the water sector is financed. With an authorized capital of RM 1 billion, a paid-up capital of RM 410 million and a nearly RM 7.7 billion worth of water assets (PAAB, 2011), PAAB clearly has the ability to assume the role which was previously the sole domain of private water operators (or international donors). Thus, the dominant role of the private sector has thus been undermined by the strengthening of the federal government, which has taken on the role of trusted financier for the sector.

This change in the financing approach triggered a shift in resource dependencies and power relations. Power relations now centre on the interactions between PAAB and state governments, rather than between the private water operators and state governments. Moreover, as a government company, PAAB is able to secure funding from both local and foreign financial markets. PAAB also benefits from being a subsidiary of the MoF, one of the prime movers behind this innovative financial approach.

This re-affirms the now dominant role of public actors in a domain previously dominated by private actors: financing. This has substantially reduced the role of private donors in the water sector. They are now (and will be) involved in just a limited number of small-scale service contracts in several states. The new public financing mechanism has done away with the situation where only private and international actors – private financers and international donors – were able to provide the needed financial resources. In one way this seems paradoxical, as limiting the role of private financing seems to be at odds with the government's pro-private sector participation, adopted under the current five-year economic development plan (Economic Planning Unit, 2010). Yet it has proven to be a model that has attracted interest from neighboring countries, hoping to learn from (and replicate) PAAB's experience in this matter.

In more general terms the reform can also be regarded as unfriendly to the business interests of the private water operators, who seem to have lost their preferential position in the water sector. Three incidences can be cited from the reform. First, the corporatization of the state water departments has dented their ambition to participate in large-scale water projects. A representative of the AUIB[45], a private water company in Kedah, claimed that they had to rewrite their business plan, since major private participation in the sector is now unlikely. This indicates

---

[45] Personal interview on 6 May 2009.

that corporatization is now favored over privatization. Second, as observed in one of the meetings (in which the researcher participated), a foreign partner in a local water consortium[46] regarded the reform as an attempt to force the exit of foreign participants from the Malaysian water market, or at least to force them to re-negotiate their concession contracts and migrate to the new licensing regime. Third, a representative of a private water operator had the impression that the central regulator 'has too much power, which could have detrimental effects on their operations'.[47] To counter the imbalance of power relations, the private water operators reacted by forming coalitions, such as the Water Association of Selangor, Kuala Lumpur and Putrajaya (SWAn) in Selangor. SWAn was created to represent the dominant role of the PNSB[48] in the Selangor water sector. In 2008, SWAn managed to gather some public support to challenge a bid by the 'new' state administration, the Pakatan Rakyat (People Pact) government, which wanted to rationalize the water sector in Selangor. At the time of writing, this situation has not been cordially resolved.

### 4.3.4 Civil society

Even though the public sector – federal and state – dominated the policy reform process, the existence of civil society should not be ignored. Their role in representing the interests of consumers and the environment was recognized. Before I discuss how they influenced the reform process, I first identify who they are. In this process three categories of civil society groups can be identified: consumer associations, environmental organizations and other interest groups.

From the initial stages, the reform attracted much attention from civil society organizations. They were drawn to the issue since water privatization was a hot political topic. They engaged in numerous platforms during the reform process, including meetings, workshops and briefing sessions. They were also given the opportunity to give feedback on the WSIA and SPANA before both legal documents were finalized. This showed a relative openness on the part of government to engage in (relatively) open consultations with civil society. However, civil society was not in a position to effectively influence the reform process. This is because some of the civil society NGOs are financially dependent on, or want to retain good relations with, state actors. Thus, public actors were able to use their resources as 'weapons' to weaken any confrontational positions of civil society towards decisions about the reform. In one interview[49] the president of the Penang Consumer Protection Association revealed that none of their comments on the major reform documents had been accepted by the government. These included calls for: the inclusion of water resources in the reform; declaring water privatization illegal; greater emphasis on the protection and conservation of water resources and to re-name the WSIA. All of these calls were dismissed. He concluded that the public platform was used as a way to 'endorse' and legitimate the reform, a view that reflected the frustrations expressed by a representative of the CAP[50]:

---

[46] French water unit Lyonnaise des Eaux together with local Pilecon Engineering hold 51% equity in the consortium.

[47] Personal interview on 6 May 2009.

[48] The PNSB is the main private water operator in Selangor, Kuala Lumpar and Putrajaya.

[49] Personal interview on 3 Aug. 2009.

[50] Personal interview on 1 Sept. 2009.

*'Our concerns were hardly considered. The participation of civil society was merely a public relations exercise (of the government). The law was drafted without serious public debate, participation and involvement.'*

However, there were several civil society organizations that did act independently from governmental influence, as these organizations were not dependent on the government for financial or other resources. The MTUC and CAWP, for example, were very vocal in their opposition to water privatization. On one occasion, together with 13 other individuals, MTUC took MEWC to the court, seeking to get the concession contracts in the Selangor water privatization made public. On 28 June 2009, the court ruled in their favour (Mei, 2009). CAWP formed a coalition with the Democratic Action Party (DAP) in pursuing its course. This coalition positioned CAWP as a force to be reckoned with. In 2008, its leader Charles Santiago contested and won a parliamentary seat in the general election on DAP's ticket. Similarly, FOMCA and MWA played influential roles, respectively championing consumer interests and showing considerable technical expertise on water. In recognition of their roles, both were appointed to the NWSC's Board of Commissioners on its establishment in 2007. These examples show that some civil society actors were able to use resources and network coalitions to increase their legitimacy and gain influence during the reform.

As the quest for greater transparency, public access to information, and efficient complaint procedures are expected to be key attributes in the water sector of tomorrow, civil society has the opportunity to further increase its influence and make its presence felt. For instance, the Water Forum[51], a public participation platform, provides an excellent platform for bringing consumer interests about tariffs and service levels to the center of the debate; the more so since the government has allowed FOMCA – the country's largest consumer coalition – to manage the Water Forum. This is a significant collaboration between state actors and civil society, with the NWSC providing financial resources and FOMCA bringing its expertise and knowledge about consumer interests.

In conclusion, the reform process demonstrated the dominant power of public actors, both at the federal and, to a lesser extent, at the state level. Both had the powers and resources required to see the reform through and influence its outcomes. By contrast private actors were marginalized during the reform process and were too weak to significantly influence its outcome. They lacked legal resources and were dependent on the states for their rights over water. Civil society's position, traditionally not very strong in water supply issues, increased significantly on consumer-related issues. Nevertheless, they are not ready (yet) to claim a major role as new guardians of the public interest, let alone to seriously challenge the legitimacy of the state in water supply (Kamat, 2004).

## 4.4 The rules of the water reform game

The water reform process involved two sets of rules. The first were the existing rules that regulated or structured the reform process. These were both formal and informal rules that existed at both the federal and state levels. The second set was the formal rules that emerged from the reform itself. It is too early to assess whether any informal rules have emerged as a consequence of the reform.

---

[51] More information can be found at www.forumair.org.my.

### 4.4.1 Formal rules structuring the reform process

The reform was introduced as a policy intervention by the federal government to improve the water supply sector. It began when a new Minister was appointed to head the MEWC in April 2004. The basis for the idea was raised during the National Water Resources Council meeting in 2003, where a decision was taken to increase the participation of the federal government in water management. This decision formed the cornerstone for the reform.

Since 1957 public policy making in Malaysia has been dominated by the procedural rules of getting Cabinet endorsement for major policies and reforms: a legacy of the Westminster style of government left by the British's colonial rule. The most important of these procedural rules is that the fundamental tenets of the (proposed) reform must be first presented to the Cabinet, the highest decision making body of the government. However, prior to that, a cabinet paper must be circulated to the related federal agencies for their comments. For the water sector reform this involved circulating the paper to the leading federal agencies: the EPU, the MoF, the AGC, the DOE and the MEWC. These federal agencies, and especially the MEWC, controlled most of the crucial resources – financial, legal, information – and so were able to determine and impose the prevailing substantive discourse that guided the policy reform. At this level of Cabinet decision-making it is normal to not directly involve non-state actors, due to the confidential nature of the subject matter. But this closed policy community can decide to allow others to give their opinions. For instance, the Cabinet consented to the request from the MEWC to conduct public hearings and to upload legal documents into the internet for public scrutiny; this later proved to be an important new precedent that emerged from the reform, as discussed in Section 4.5.3.

The procedural rules of state governments were also crucial in the final stages of the reform process. The states also have cabinets, known as the State Executive Councils (EXCO), headed by a Chief Minister. In addition every state is headed by a Ruler. These two formal institutions play an important role in the state administration. The cooperation of state governments was needed for the reform to run smoothly, especially after water was moved from the State List to the Concurrent List in 2005, necessitating shared responsibility and decision-making. Hence, during the negotiation process of the water reform there were constant interactions and often joint decision making between the federal agencies involved in the water reform and the main state governmental bodies, the EXCOs and the Rulers. At least two examples can be used to illustrate this. First, the corporatization of the state water department could not be smoothly implemented and could even be severely delayed without the consent of the EXCOs and/or Rulers. Without such consent the policy paper for corporatization could not be presented to the state assemblies. Second, in implementing the water sector reform State Water Acts had to be amended or new acts had to be formulated. Only the State Assembly has the power to effect both kinds of actions. The Federal state was well aware of the need to respect this and its dependency on the co-operation of state governments in moving forward with the water sector reform. During the course of the reform, the state governments were constantly consulted by the federal agencies about matters touching on the states' interests, and this included a set of public hearings processes (held from 2005 to 2006) as well as informal bilateral briefings and lobbying. The states' Rulers were formally briefed in two separate meetings. Here we see that the reform process was regulated by quite a number of formal procedural steps and decision making mechanisms at both the federal and state level.

### 4.4.2 Informal rules structuring the reform process

These formal rules structuring the interactions of state bodies at and between federal and state levels were accompanied by frequent informal interactions between federal agencies and between federal agencies and state level bodies designed to arrive at common positions and a shared agenda. Many such meetings took place. These informal contacts ensured that such a major reform, in which powers (over natural resources, over finances, over infrastructures) were reallocated between state and federal levels and between public and (semi-) private bodies, progressed quite smoothly and without major conflicts. Such informal interactions and information sharing became an unwritten but influential rule within the rather closed policy community, with predominantly public state actors.

The first informal rule involved the unprecedented involvement of the Prime Minister and his Deputy in the reform process as highlighted in the previous section. Their presence in three meetings smoothened the process and mitigated the conflicting interests of federal government actors and, to some extent, state government actors. Under normal policy-making process, ministries table their intention to formulate policy in the weekly Cabinet meetings. Through this platform (and the three unprecedented meetings) both men were directly informed about policy concerns. This direct involvement of both top governmental officials in the policy reform was exceptional. It did not follow any normal written/formal rule of policy making. As this hardly ever occurs, it indicates that the (direct) involvement of both men was needed to guide a policy issue as controversial and complex as water. We have seen that the reform was about far more than just (improving) water supply. It also had implications for the sovereignty of state governments and these two men were obviously in a better position than anybody else to convince state governments to consent to and collaborate on these reform proposals. It is hard to imagine that both men would be so readily involved in policy making processes of a less controversial nature, such as transportation or communications.

There were many other innovative informal rules that guided the water sector reform, particularly those involving interactions between state and non-state actors. As stated earlier, one of the unprecedented moves taken by the government was to organize public hearings, a form of participative (or at least consultative) policy making. This informal rule allowed greater public participation in the policy reform. From late 2005 until early 2006, over 50 sessions were held to seek feedback particularly on the two legal documents, involving a wide range of stakeholders in the water sector and other interested parties. The government even took a bold step of engaging politicians from the ranks of both the government and the opposition. This was the largest public consultation exercise ever undertaken in the history of the country. The willingness of the government to engage the public, the private sector and the opposition in policy making can be interpreted as setting a precedent for more open, transparent and consultative approaches.

This platform allowed non-state actors, especially private water operators and civil society, to participate in the discussions on the reform. The motivations for joining these public hearings were diverse: some were moved by the desire to demonstrate and protect their interests (as in the case of the PNSB in Selangor), others participated to object to the reform (as in the case of Lyonnaise des Eaux in Johor). Civil society organizations used the platform to urge the government to ban water privatization and to adopt public-public partnership (as in the case of CAWP), or to demand the inclusion of water resources in the reform (as in the case of CAP). Even though these groups did not succeed in influencing the core of the decisions of the reform, their right to play an active role was recognized, the water reform was legitimized and a new practice of participation entered public policy making processes.

It is not by accident that public hearings of such magnitude happened on the policy issue of water supply and its reform. Water reform touches upon socially, economically and politically sensitive matters. Socially, (access to) water is a human rights issue, with the government having an obligation to ensure that every citizen has access to water. The reform also touched upon essential aspects of federal-state relations. Economically, water is important: the prices of such essential goods are always very sensitive, yet before reform little regard was paid to the economic value of water. Subsidized tariffs led to over consumption and the wastage of water, but major water price increases could equally result in unrest if not managed properly. Politicians realized that there was much political mileage in water issues and these were an important concern for the masses. The public hearings reflected the desire of the government to address the controversial and sensitive nature of these issues and to increase public engagement in policy making. While the government did not actively promote public engagement in these hearings this change in political culture may well pave the way for a wider application of participative policy making in the future. In developed countries, participative policy making has gained increased acceptance not only because it enhances democracy but also for its potential to contribute to effective, efficient and accountable systems of governance (Zhong, 2007). Since the water reform this approach has made a further inroad into Malaysian policy making process, in discussions over the Personal Data Protection Act in 2010. In that sense, we might speak of the institutionalization, and perhaps even formalization of an initially informal rule in policy making.

The second informal rule of the game that affected and structured relations between state and non-state actors in the reform process was the declassification of (confidential) public documents to allow public scrutiny. Usually, draft legal documents are regarded as highly confidential and protected under the 1972 Official Secrets Act. These documents can only be publicly disclosed after being tabled to, and approved by, Parliament, after they have become law. However, drafts of both the WSIA and SPANA were made available to the public through MEWC's website for one month. This step was taken to facilitate the public hearings and soliciting stakeholder feedback on both documents.

Another difference from conventional policy making was the use of (then) new media – the Internet – for its speed, convenience, access and potential of reaching a wider audience. As expected, the response on both documents was overwhelming. More important than this, the use of new media reflected (a degree of) political readiness by the government to embrace innovations in policy making processes. Making documents public involves disclosing information, upholds the 'right to know' of the citizens and involves inviting them to be part of the policy process. These goals were a further aspect of the aspirations of water sector reform, which promoted informational governance (Mol, 2008) in the water sector (Section 29 of the WSIA). In this respect information technology (or e-governance) played a key in helping non-governmental actors to gain wider access to public documents, which traditionally had remained closed within a narrow governmental sphere/policy community. This helped to narrow the information asymmetry between governmental and private actors and allowed non-governmental actors to use this information as a resource to try to influence the reform process. At the time of writing, increased transparency and disclosure of public documents has not been repeated and it remains unclear if the government will consider adopting this practice (either formally or informally) in future policy making.

### 4.4.3 Rules that emerged from the reforms

The reform emphasized the need for formal arrangements of engaging and promoting active public participation in the water sector. This shift to including the public in policy making was manifest in the institutionalization of a formal institution known as the Water Forum, established under Section 69 of the WSIA (MEWC, 2006). The Water Forum was specifically created to represent the public, especially as consumers, in the water supply sector (Page & Bakker, 2005). Its remit was mostly limited to matters concerning consumer interests, especially tariffs and service quality (as clearly specified under Section 70 of WSIA). With this focus it comes as no surprise that FOMCA – the largest national coalition of consumer associations – was entrusted to manage Water Forum on behalf of the NWSC. This platform gave the public an opportunity to raise their concerns and influence the reform process. From another angle, the Water Forum could, in the future, become the 'eyes and ears' of the public, complementing the (regulatory) role of the NWSC. Section 70 (d) allows the Water Forum 'to identify and keep under review matters affecting the interests of consumers and ensure that the water supply services and sewerage services companies are aware of, and responsive to, concerns about their services' (MEWC, 2006a: 62). Civil society organizations believed that the intended objective of Water Forum could be compromised if it became dominated by market actors, especially water companies.[52] These organizations demanded that the word 'industry' would be removed from Section 69 of WSIA, in which they succeeded. While the presence of this word might have suggested that the water companies could exert some influence on the forum, membership of the Water Forum is open to everybody, so there is nothing to stop water company representatives from participating (in a private capacity) in the forum. At the time of writing, the water companies have not made any inroads into the Water Forum. It is expected that the NWSC will take steps to ensure that the Water Forum remains a platform for representing public and consumer interests in the water sector, rather than those of private water companies.

Another important output of the reform was the clear separation between ownership, regulation and service provision. Traditionally, these three functions were performed by the state governments. This new division of functions between different bodies was built on the premise that dispersing and assigning them to different bodies would minimize the likelihood of conflicts of interests emerging and would also strengthen regulation, both considered essential in developing an efficient and effective water supply sector. Weak and ineffective regulation has been cited as (one of the) contributing factors to the failures of water privatization around the world (e.g. Casarin *et al.*, 2007). Several institutional arrangements were created during the Malaysian water reform to prevent this from occurring. For instance, a central regulatory body and state water corporations were established to separate ownership, regulation and service provision. Even though this practice,

---

[52] Interviews with various civil society organizations.

which promotes good governance, has not yet been widely adopted in other sectors, it has been replicated for solid waste management, where policy formulation has been allocated to the Solid Waste Management Department, monitoring, enforcement and control to the Solid Waste and Public Cleaning Corporation, with the private sector providing the services. Both governmental bodies were established in 2008 (The Solid Waste and Public Cleaning Corporation, 2010).

The third formal rule that emerged from the reform concerns a new approach in water infrastructure financing. The previous financial arrangements were highly dependent on federal budgets allocated under the five-year development plan (known as the Malaysia Plan). The new rule introduced an innovative mechanism for financing the water supply sector, leveraging the strengths of both the public and private sector, akin to a kind of public-private partnership. It is envisaged that this rule will cushion the effects of (costly) private financing through tariffs and circumvent the water utilities being over-dependent on limited federal financial resources for water infrastructure development. A public-owned institution (PAAB) was established in 2006 to translate this rule into practice. PAAB is entrusted to consolidate the ownership of the water infrastructure under a single body, and be responsible for maintaining a separation between water provision and ownership of assets (MEWC, 2008). In business terms, PAAB is a single purpose vehicle created to provide the financial parts of water sector or any other public sector transformation. On a practical level, this rule operates on two stages. At one level it involved a transfer of assets, with PAAB taking over the state government's outstanding debts in exchange for water assets. On another, PAAB operates build and lease schemes, acquiring the funds to build water infrastructure and leasing these to water operators on completion, for an agreed rent. To avoid unnecessarily high tariffs, an affordability principle is used to determine the lease rental. As such the lease rental may differ from one company to the other. The objective of this rule is to facilitate all water operators and the PAAB to achieve full cost recovery and financial independence.

This new financial arrangement re-affirms PAAB as the dominant force in water infrastructure financing, a role previously held by either the federal government or the private water operators. Backed by relatively rich resources – financial and water assets – and powers gained as a federal body, there is no doubt that PAAB will become the biggest water infrastructure financier in the country. Six state governments have already adopted this approach and similar arrangements are being made with others, indicating its effectiveness and dominance. No other financial arrangement can compete with those offered by PAAB, which are specifically designed to take into account the long-life span (30-50 years) of water assets (MEWC, 2008).

## 4.5 Conclusions

This analysis of changing policy arrangements has produced useful insights into the discourses, actors, powers and rules that came along the reform of the Malaysian water sector and that emerged from that process.

Prior to the reform, the water supply sector was firmly located within the domain of state governments, with no clear division between the responsibilities of ownership, regulation and service provision. In several states, the private sector was strongly involved in service provision, while civil society was hardly involved, other than as a captive consumer.

The reform reorganized the water supply sector along the principle of shared-responsibility. Specific tasks were delegated to more parties creating a landscape of multiple actors working at multiple levels (Figure 4.1). In that sense, the water reform is a product of, and further proves the interpretative powers of, modern governance theories. The federal government took responsibility for regulatory oversight, while the role of state governments in water provision was further strengthened through corporatization and practices, and as owners of water resources. A new para-statal unit (PAAB) was formed to take on the financial responsibility, thus reducing the role of the private sector and international donors in water financing. The corporatization discourse has reduced the chances for the private sector to participate in (full) privatization, while it has allowed state water departments to profit from a more business-like culture injected into public water provisioning. The presence of civil society became more relevant, with the creation of formal public participation through a platform known as the Water Forum. The existence of relatively strong and effective non-state actors in civil society partially balances the dominant power of state actors. Nevertheless, these civil society actors are not (yet) ready to claim a role as the new patrons of public interest, let alone pose a serious challenge to the dominant role of the state. As such, it can be concluded that the water sector reform, strengthened the role of the state (and especially the federal state) in the water sector.

The process was guided and regulated by a set of formal and informal rules and in turn established new formal rules and precedents. Both these sets of rules served to institutionalize the dominance of state actors, (at both federal and state governmental levels), increased the role of civil society and reduced the role of the private sector.

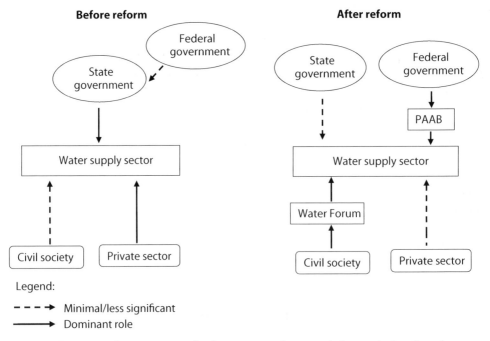

*Figure 4.1. Institutional arrangements for the water supply sector: before and after the reform.*

# Chapter 5.
# Operational efficiency of the water supply sector

## 5.1 Introduction

Spiller and Savedoff (1997) regard the urban water sector in many developing countries as having a 'low-level equilibrium', or low operational efficiency. As a branch of public services, the water sector is often labelled as inefficient and unable to meet the rapidly growing demand (Bhuiyan & Amagoh, 2011). As a consequence, the World Bank views water utilities, particularly in developing countries, as being 'locked in a vicious cycle' of weak performance incentives, low willingness of customers to pay cost recovery tariffs, and insufficient funding for maintenance, ultimately leading to a deterioration of assets and a squandering of financial resources (Hayward, 2007; Stedman, 2009).

There is also pressure for the water sector to support increases in economy growth as countries develop. This has resulted in many (developing) countries reforming their public water sector to break this 'vicious cycle' and to meet the growing demand for water created by economic growth. This has led to the traditional roles of the government as a water supplier and a regulator being decentralized and new rules facilitating greater participation of market actors to be introduced (Asian Development Bank, 2006). As a result, new management styles have emerged: New Public Management, corporatized-public, public-private partnership (PPP) and private operations (Rouse, 2007; Schwartz, 2008). Among these, PPP and privatization have attracted fierce criticism from scholars such as Holland (2007) and Kessides (2004) who question the supposed efficiency gains of privatization in the water sector.

In the case of Malaysia, the water sector had been suffering from several long-standing operational issues: of high levels of non-revenue water, insufficient revenue due to below-cost recovery tariffs, inefficient management and an absence of an (effective) customer complaints management system. The water sector reform aimed to address these issues, and this chapter analyzes in detail how effectively this goal has been met. Wherever possible this analysis compares the period before and after the reform. However, due to data scarcity; such a comparison was not always possible. Section 5.2 briefly describes the methodological approach to the research. Section 5.3 presents the assessment of four operational efficiency indicators: non-revenue water, collection efficiency, unit production cost and customer complaints. The analysis is deepened by two in-depth cases studies – one of PBAPP, and one of SADA – in Section 5.4. The chapter ends by discussing the conclusions (Section 5.5).

## 5.2 Methodological approach to the research

### 5.2.1 Sources of data

Data for this chapter were generated through three main sources: secondary data collection, in-depth interviews with key informants and surveys.

The main source for the secondary data for this chapter was the Malaysia Water Industry Guide (MWIG) from the years 2005 to 2011. The MWIG is the Malaysian Water Association's (MWA) yearly publication of indicators of the performance of water utilities. This source provided data for non-revenue water, unit production cost and customer service complaints. Semi-structured questionnaires were used to gather data for collection efficiency from water utilities.

1. *Non-revenue water (NRW)*. The MWIG presents a general set of data, which indicate the percentage of NRW at the state level. However no further details about the components of NRW are reported. This is in contrast to the recommendation of the International Water Association (IWA) and the widespread practice of breaking NRW data down into two major components: physical and commercial losses (Farley *et al.*, 2008).

2. *Unit production cost*. There are two components of production cost. The first component represents the basic costs associated with the treatment and distribution of water. The second component also includes the financial costs – capital expenditure, depreciation and other financial costs. Since the MWIG only reports the basic costs, these (and not the financial costs) are used for the analysis.

3. *Customer service complaints*. Data from the MWIG contain two components: number of interruptions to the service per year, and other complaints[53] (MWA, 2010). The American Water Works Association suggests that complaints can be categorized into (a) service-related complaints, and (b) technical complaints (AWWA, 2004). Data for customer complaints are best presented in relative figures: the number of complaints per 1000 connections. Since data gathered from the MWIG are in absolute figures, manual calculations were performed to obtain the number of complaints per 1000 connections (i.e. number of complaints/thousand connections).

4. *Collection efficiency*. The MWIG does not collect data on collection efficiency. To cover this information gap questionnaires were sent out to twelve water distribution utilities, although only four – Syabas, SADA, SAJH and PBAPP – returned the questionnaires.

Other secondary data were sourced from water utilities' annual reports[54] and government publications, such as the 9th and 10th Malaysia Plan, the National Water Resources Study, etc.

Finally, in-depth interviews were conducted with various key informants and stakeholders in the sector. These included: water operators, NGOs (consumer associations and environmental organizations), and government officials. Interviews were also conducted with the top management of PBAPP and SADA for the case studies (seven interviews with PBAPP and six with SADA). Further information was gathered through the personal contacts that the researcher has with the management of both companies and other stakeholders in the water sector, through observations and site visits.

---

[53] Represent illegal connections, billing disputes, water quality, low pressure, etc.

[54] Applicable only to water utilities listed in the Malaysia Stock Exchange.

## 5.2.2 Data analysis

The analysis is conducted on the basis on ownership – public and private – to compare the effect of the reform on the operational functions of the two types of company. Public water utilities are further categorized into (1) state water departments; and (2) corporatized entities (Table 5.1). The water utilities in the states of Sabah and Sarawak were excluded from the analysis since these states decided to stay out of the reform and were unaffected by the process (Hua, 2009).

Some further explanations should be given about these categories. First, the corporatization of PWD Perlis was approved in 2010 but the implementation has not yet gone through. Thus, PWD Perlis is still effectively a state water department. Second, SAMB, SAINS and SADA only came into existence in 2006, 2009 and 2010 respectively, when state water departments were corporatized. Prior to that, they were state water departments, known as JBANS, PAM and JBA Kedah respectively. Data before these dates refer to the time when these companies were operating as state water departments. Third and lastly, Syabas only came into existence in 2005 when water provisioning in the state of Selangor was privatized. Thus, data for 2004 relate to PUAS, the state water corporation at the time.

*Table 5.1. Malaysian water utilities by category (MWA, 2009).*

| Category | Water utilities | Area served |
|---|---|---|
| State water department | JBA Labuan | Labuan |
| | JBA Pahang | Pahang |
| | PAM[1] | Melaka |
| | JBA N. Sembilan[2] | N. Sembilan |
| | JBA Kedah[3] | Kedah |
| | PWD Perlis | Perlis |
| Corporatized water department | AKSB | Kelantan |
| | SATU | Terengganu |
| | LAP | Perak |
| | SAMB[1] | Melaka |
| | SAINS[2] | N. Sembilan |
| | SADA[3] | Kedah |
| Private entity | SAJH | Johor |
| | Syabas | Selangor[4] |
| | PBAPP | Penang |

[1] PAM remained as the state water department until 1 July 2006 before it was corporatized into SAMB.
[2] JBA N. Sembilan was corporatized into SAINS on 1 July 2006.
[3] SADA came into existence on 1 Oct. 2010 when JBA Kedah was corporatized.
[4] Includes Kuala Lumpur and Putrajaya.

## 5.3 The operational efficiency of the Malaysian water sector

### 5.3.1 Non-revenue water: the art of managing leaks

*Overall performance*

NRW refers to water lost in the distribution system before it can be sold to customers to generate revenue (Winarni, 2009). From a technical perspective, high (and increasing) water losses reflect 'ineffective planning and construction, and of low operational maintenance activities' (Lambert, Brown, Takizawa & Weimer, 1999: 227). High water losses adversely affect the revenue of water utilities and imply higher per unit production costs (Stedman, 2009). Worldwide, it has been estimated that water utilities suffered US$ 14 billion in financial losses through leakages (Farley *et al.*, 2008).

NRW in Malaysia declined by, on an average of 3% between 2004 and 2009, an average yearly reduction of a mere 0.5% (Figure 5.1). The highest drop of 2.1%, recorded in 2004-2005, can be attributed to an increase of 73% in budgets for water infrastructure under the 8th Malaysia Plan (2001-2005) (Economic Planning Unit, 2008). This allowed water utilities to undertake remedial works to reduce losses mainly resulting from pipe leakages and bursts, the main cause of losses. As a result of a further 78% increase in budgetary allocation in the 9th Malaysia Plan (2006-2010), there was a further reduction in NRW between 2005 and 2009 (Economic Planning Unit, 2008). However, with a current rate of just 0.5% per annum reduction, it will still take another 33 years before reaching the 20% target set under the reform (MEWC, 2008; MWA, 2010). This target appears particularly ambitious considering that, in 2010, a drop of just 0.2% was recorded (MWA, 2011). Nevertheless, the figures show that major investments do contribute to reducing these losses.

*Performance by state water departments*

NRW performance re-affirms the operational inefficiency of public water utilities. JBANS, JBA Pahang and JBA Kedah, all public utilities, suffered huge water losses in the range of 45% to 60% (Table 5.2). One of the prime objectives of the reform was to reduce these unsustainable

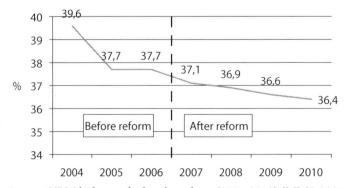

*Figure 5.1. Average NRW before and after the reform (2004-2010) (MWA 2004-2011).*

*Table 5.2. The NRW of state water departments (2004-2010) (in %) (MWA, 2004-2011).*

| State | Water utilities | 2004 | 2005 | 2006 | 2007 | 2008 | 2009 | 2010 |
|---|---|---|---|---|---|---|---|---|
| Kedah | JBA Kedah/SADA | 42.8 | 43.8 | 45 | 41.7 | 44.9 | 44.9 | 43.0 |
| Labuan | JBA Labuan | 28.7 | 24 | 36 | 35.9 | 33.2 | 25.8 | 24.9 |
| Melaka | PAM/SAMB | 33.4 | 29.8 | 27 | 29.8 | 30.1 | 29.7 | 26.0 |
| N. Sembilan | JBANS/SAINS | 54.7 | 53 | 60.1 | 53.8 | 50.5 | 49.2 | 43.4 |
| Pahang | JBA Pahang | 48.2 | 49.7 | 46.4 | 53.6 | 52.9 | 59.9 | 55.3 |
| Perlis | PWD Perlis | 36.7 | 36.3 | 35.5 | 34.1 | 41.7 | 44.7 | 51.3 |

rates of NRW. The reform motivated the state governments of Negeri Sembilan and Kedah to re-examine the effectiveness of their water departments, which resulted in JBANS and JBA Kedah being converted into business entities through the process of corporatization in 2009 and 2010 respectively (Rouse, 2007; Fuest & Haffner, 2007). JBANS successor, SAINS, recorded a significant reduction in NRW of almost 6% in 2010, its second year of operation (MWA, 2011). Even though there is not yet any clear evidence to link the reform to reductions in NRW, we can make the preliminary conclusion that the 'new private sector working culture' has the potential to facilitate greater efficiency in (public) water utilities. This said, corporatization has yet to improve the level of efficiency for SADA (the successor to JBA Kedah), which is assessed in detail in Section 5.4.

The delay in the corporatization of JBA Pahang[55] may well be connected with a general increase in NRW levels in the state. By contrast, JBA Labuan and PAM/SAMB have achieved reductions in their levels of NRW in recent years. This said there are some questions about the reliability of these data. The secretary-general of MWA is on record as saying that 'public water utilities rely on desk top data, rather than data from a valid source. Data from private water utilities are more reliable'.[56] This raises questions about how JBA Labuan and PAM/SAMB (and many others) derived their NRW data. Reliable data is crucial for determining NRW figures and trends.

High NRW has caused huge financial losses for state water departments. In 2009, for instance, JBA Pahang and SAINS suffered water losses equivalent to 200,127 m$^3$ and 118,248 m$^3$ respectively (MWA, 2010). At the average national tariff of RM 0.92/m$^3$, the cumulative financial losses suffered amounted to RM 292,905. This severely affected the financial stability of both utilities. Even though both utilities managed to reduce the amount of losses in 2010, only SAINS translated this into a sharp reduction in their deficit, and it did not have significant impact on JBA Pahang's deficit (Figure 5.2) (MWA, 2009, 2010).

What then have been the main obstacles preventing state water departments from effectively addressing the issue of NRW? Evidence obtained from the interviews puts the inability of state water departments to address NRW down to two main factors: funding and management issues.

---

[55] Expected only to happen in 2012.

[56] Personal interview on 3 March 2011.

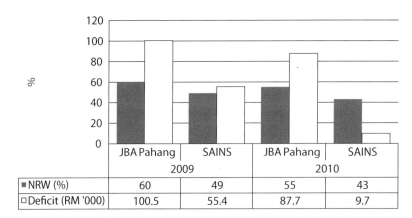

Figure 5.2. The effect of NRW on the financial performance of JBA Pahang and SAINS (2009-2010) (MWA 2009-2011).

It is worth asking why, despite the increase in water development budgets, only minimal reductions in NRW have been achieved in recent years. Under the 9th Malaysia Plan the budget increased by 73%, but the reduction in NRW was minimal: just 1.1% (Economic Planning Unit, 2008). Two conclusions can be drawn from this. First, as indicated by state water departments, such as JBA Pahang, SAINS and JBA Kedah, the increase in budgets was far from sufficient to undertake effective NRW reduction programmes (MEWC, 2008). Moreover, the budgets were spread over a 5 year period and shared amongst 14 states. To a certain extent, insufficient funds forced state water utilities to make difficult choices about NRW projects, and more often than not, pipe replacement took priority. According to the Director of the Planning, Coordinating and Monitoring Division from the Water Supply Department of the MEWC, pipe replacement alone was not enough to address the NRW problems. He thought that 'the main problem is that the NRW programme is not implemented holistically. In some states, the focus is on changing pipes rather than addressing problems in totality'.[57] A representative from the MWA shared this concern, arguing that NRW involves more than just pipe replacement. It involves 'allocating resources to set up the District Metering Areas (DMAs), mapping all of the piped system and undertaking active leakage control to effectively reduce losses'.[58]

Second, this lack of progress not only reflects an inefficient use of money, but also the poor execution of plans and programmes to reduce NRW by the states. The secretary-general of the MWA observed that 'it is shocking when the Audit Department reveals that, in some states, the NRW budgets are invested in a place where there is no water'.[59] This situation forced the federal government to explore a more effective (and sustainable) way to facilitate project financing and management in order to obtain better value for money. Starting from the 10th Malaysia Plan (2011-2015), financial assistance to state governments for water infrastructure will gradually be

---

[57] Personal interview on 4 March 2011.

[58] Personal interview on 3 March 2011.

[59] Personal interview on 3 March 2011.

phased out (except for dams) and PAAB[60], the government's water asset management company, will become responsible to raising funds for, and overseeing the overall planning and implementation, of water infrastructure projects.

Unlike their counterparts in the private sector, state water departments did not have a dedicated team or department to handle NRW related-issues, partly due to a lack of know-how and partly because of budgetary constraints. Existing (rigid) recruitment procedures and budget constraints meant that the state water departments have been unable to attract the skilled workforces needed to expand the network or improve their performance. In general they engaged outside contractors to implement and monitor NRW reduction programmes. Budgetary constraints often forced state water departments to limit the outsourcing to 'implementation' with no provision available for 'monitoring'. Problems started to reappear when the project was handed back to the water utilities which should have had a dedicated team to continue and monitor the projects. According to a senior engineer from the Water Supply Division, NRW management 'requires a mental change among state water utilities so they see the management of NRW as a (continuous) programme and not on a project basis and the importance of continuous monitoring'.[61] If a water system is left unattended or unmonitored after the hand-over of a project leakages of some form will inevitably occur and continue to increase each year. This phenomenon is often referred to as the natural rate of rise of leakage (McKenzie & Lambert, 2002). In Malaysia, the MWA has estimated the average natural rate of rise of leakage to be around 1.3% per year.[62]

## Performance by corporatized water entities

Two of these entities (AKSB and SATU) recorded an increase in the level of NRW in the period after the reform, while LAP maintained the level of NRW at more or less stable levels across both periods (Table 5.3).

The situation faced by AKSB and SATU was contributed to by a lack of investments in NRW programmes due to budget constraints. By contrast continuous monitoring and an on-going implementation of NRW programmes helped LAP to record one of the lowest NRW rates in the country.

*Table 5.3. The NRW of corporatized water entities (2004-2010) (MWA, 2004-2011).*

| State | Water utilities | 2004 | 2005 | 2006 | 2007 | 2008 | 2009 | 2010 |
|-------|-----------------|------|------|------|------|------|------|------|
| Kelantan | AKSB | 40.8 | 40 | 44.4 | 48.4 | 49.4 | 48.3 | 52.4 |
| Perak | LAP | 31.7 | 30.6 | 30.7 | 30.1 | 31.2 | 30.7 | 29.4 |
| Terengganu | SATU | 33.3 | 34.7 | 31.5 | 38.5 | 38 | 37.9 | 39.4 |

---

[60] PAAB was established on 5 May 2006.

[61] Personal interview on 4 March 2011.

[62] Personal interview on 3 March 2011. This is based on what is experienced by SAJH.

AKSB is the water utility owned by the state government of Kelantan, one of the states governed by the opposition party, the Parti Islam Malaysia (PAS). Federal budgets of RM 60 million for water infrastructure development to AKSB were withdrawn after PAS gained control of the state administration from a coalition government, the Barisan Nasional (BN) at the 1990 general election. Only during the 8th Malaysia Plan (2001-2005), was federal assistance for water infrastructure restored to the state. In that period, the state was given RM 261 million for water works; far less than the RM 600 million it requested. At that time most of the state's water infrastructure was already in a poor state (Sahabat Alam Malaysia, 2004). Under the 9th Malaysia Plan (2006-2010), the budget to the state was reduced to RM 135 million. Even so there were problems with the implementation of the planned water projects which were not completed at the end of the plan period (end 2010). This, to certain extent, jeopardized the programme to replace 3,632 km of leak-prone asbestos cement pipes old communication pipes and over 30,774 un-economic water metres (MWA, 2010).[63] Thus, certain projects – such as the NRW projects for the district of Kota Bahru – have been carried forward into the 10th Malaysia Plan (2011-2015).

Lack of investment in water development infrastructure also hampered SATU's ability to effectively contain water losses. For instance, in 2009 alone SATU needed to invest heavily to replace more than 1,531 km of leak-prone asbestos cement pipes and to change 95,497 old water meters to avoid incorrect readings (MWA, 2010). However, SATU only got RM 46 million to do this under the 9th Malaysia Plan (Economic Planning Unit, 2008): 76% of what it had asked for. As a result, its NRW level increased to 39.4% in 2010 – far above the national average (MWA, 2010). High water losses also caused huge damage to SATU's financial standing, with its deficit increasing almost six fold: from RM 1.7 million in 2008 to RM 6.1 million in 2009. It recorded even a bigger deficit in 2010 (MWA, 2011). At the average tariff of RM 0.84 per $m^3$, SATU suffered financial losses that amounted to RM 72 million in that year alone. Operating with such a deficit has clearly not helped SATU to allocate sufficient funds to tackle its problems with NRW. The evidence lead to us to conclude that, at this point of time, the reform has not helped SATU to manage its NRW efficiently and it is reasonable to say that the reform is unlikely to have an immediate effect on the company's ability to improve its NRW losses. However, when SATU agrees to fully migrate and subscribe to the reform particularly regarding the financial arrangement (by PAAB), we can expect that SATU would financially benefit in terms of accelerating the NRW projects – including pipe replacements, pressure management, establishing District Metering Zones (DMZs) and changing meters – all of which are core elements of successfully reducing water losses.

Judging from the level of NRW it had achieved, LAP was one of the most efficient water utilities: second only to PBAPP. It managed to get its NRW levels well below the national average (MWA, 2010, 2011). According to the General Manager, the management of LAP treated its NRW reduction programmes as an on-going process rather than as one-off assignments, and continuously monitored them.[64] In fact, the Water Supply Division of the MEWC commended LAP's commitment and passion in addressing NRW and particularly the establishment of a dedicated company, within LAP, to address NRW in the state.[65] Years of hard work in reducing

---

[63] 7 years old and above.

[64] Personal interview on 22 May 2009.

[65] Personal interview on 4 March 2011.

NRW has also bought financial rewards for LAP. It recorded a surplus of RM 74.4 million in 2009, an increase of 33% from RM 49.6 million achieved in 2008 (MWA, 2010). In 2010, LAP recorded even a higher surplus of RM 90.6 million (MWA, 2011). This goes to show that effective NRW management is not just about investment, but also about follow-up and monitoring both of which are vital in preventing the natural rate of rise of leakage from (re)occurring.

## Performance by private water entities

As anticipated, private water utilities demonstrated a better NRW performance than their counterparts in the public sector. PBAPP recorded the lowest NRW of any water utility, while SAJH[66] and Syabas[67] managed to significantly reduce their NRW between 2004 and 2010 (Table 5.4). Overall, the NRW levels in the private sector also fluctuated much less over the years (Table 5.4). Compared to state-run water utilities, SAJH and Syabas were financially able to undertake effective NRW reduction programmes. A separate assessment for PBAPP is presented in Section 5.4.

SAJH and Syabas are only involved in downstream activities – distribution, billing and collection – and purchase treated water from companies that treat the water.[68] As private entities, there are financial incentives to reduce NRW (to specified levels) set out in their concession agreements. For example, Syabas is required to reduce its NRW from 43% to 15% by 2035 (Syabas, 2005). Meanwhile, SAJH was obliged to reduce its NRW to a level of 20% level by 2010[69] (Ranhill Utilities Berhad, 2006). The reward for meeting these targets is a permitted tariff increase. However, Table 5.4 shows that SAJH has failed to meet that target, while it is (still) not yet possible to formulate conclusions in the case of Syabas.

Since taking over water distribution operation in the state of Johor in March 2000, SAJH managed to reduce NRW from 36% in 2004 to 30% in 2010. Since its establishment in December 2005, Syabas managed to reduce NRW in Selangor from 38% to 32% in 2010. One of the main

Table 5.4. The NRW of private water utilities (2004-2010) (in %) (MWA, 2004-2011).

| State | Water utilities | 2004 | 2005 | 2006 | 2007 | 2008 | 2009 | 2010 |
| --- | --- | --- | --- | --- | --- | --- | --- | --- |
| Johor | SAJH | 36.3 | 35.5 | 32.5 | 31.2 | 31.3 | 31.9 | 29.9 |
| Penang | PBAPP | 21.4 | 19.4 | 18.6 | 16.8 | 16.9 | 19.1 | 18.2 |
| Selangor | Syabas | 42.7 | 38.4 | 36.6 | 34.7 | 33.9 | 32.5 | 32.5 |

---

[66] The SAJH entered into the concession with the state government of Johor on March 2000 under Johor State Water Supply Privatization Scheme.

[67] Began operation on December 2005.

[68] The SAJH purchases treated water from two treatment operators: the Southern Water Corporation and Equiventures. Syabas purchases treated water from three treatment operators: Puncak Niaga, Konsortium ABASS and Splash.

[69] Refer to Table 5.4 for actual NRW achievement.

contributors to this reduction was the huge financial investment agreed upon under the concession agreement with the state government. During the entire concession period, SAJH and Syabas are obliged to invest RM 1.6 billion (Ranhill Utilities Berhad, 2009) and RM 10.7 billion respectively in capital works, mainly on NRW projects (Syabas, 2005). Statistics obtained through a survey indicated that in 2008 alone, Syabas and SAJH invested RM 286.5 million and RM 17.7 million respectively in various NRW reduction programmes.

Besides focusing on pipe replacement and changing water meters, both SAJH and Syabas are convinced of the need to formulate an effective NRW management system. In 2004, SAJH formulated a NRW Strategy and Action Plan. Within this framework, it introduced a Job Management System, Remote Sensing System and DMZs for efficiently monitoring and managing its NRW initiatives (Ranhill Utilities Berhad, 2008). Syabas has also implemented DMZs, Pressure Management Zones and replacing electromagnetic flow meters to reduce water losses in Selangor, Federal Territory of Kuala Lumpur and Putrajaya (PNHB, 2009). However, an on-going dispute between the company and the state of Selangor about tariff revisions for 2009 has affected Syabas's capability to accelerate its NRW reduction programmes.

### 5.3.2 Collection efficiency: ensuring what is billed is collected

Most of the water utilities bill their customers for water consumption on a monthly basis. As water utilities derive their revenues from the sale of water, it is crucial that every bill that is issued is paid and accounted for as this determines how much revenue the water utilities receive. In short, water utilities must strive to attain the highest possible ratio between the amounts of revenue collected against the bills issued. The higher this ratio is, the higher the revenues to the water utilities.

One can't deny the importance of water tariffs to water utilities. IWA's Executive Director, was quoted in the June 2009 edition of Water Utility Management International as saying that, at a very minimum, a water tariff has to recover the cost of the services. However, he lamented that, even in developed nations, authorities are 'afraid to lift rates to reflect what the system needs' and these remain below the full cost of supply (Stedman, 2009; Rogers, de Silva & Bhatia, 2002). Many governments, developed and developing alike, want to ensure that water tariffs are kept low and affordable to improve citizens' access to water (Hua, 2009). Reddy (1998 in Crase & Gandhi, 2009) argued that restructuring tariffs with the aim of achieving cost recovery was itself a formidable task – not just a technical one, but also involving institutional challenges. It is about finding solutions as to how 'water supply be organized and financed; and how institutions can develop better incentives and make improvement more sustainable' (Asian Development Bank, 2006: 37).

In the 2008 World Water Congress held in Vienna, the IWA reported that globally only 31% of water utilities achieve full cost recovery. These utilities were mainly found in developed countries (Hua, 2009). In Malaysia, in 2009 water tariffs covered only 78% of the operating expenditure (Economic Planning Unit, 2010). To be sustainable, 'tariffs should reflect the full cost of water, including full supply cost, opportunity cost, economic externalities and environmental externalities' (Pearce-Oroz, 2006; USAID, 2005). Having to operate in the environment where the tariff is below this level poses a great challenge to many water utilities. To maintain robust revenue streams, probably the best they can do is to improve their revenue collection efficiency, and decrease the number of uncollected bills.

*Performance by ownership*

As explained earlier, data on collection efficiency are not presented in the Malaysian Water Industry Guide. Hence, questionnaires were used to gather data about this indicator. 12 questionnaires were distributed but only four water utilities – SAJH, Syabas, PBAPP and SADA – returned them. The first three of these are private water utilities, while SADA is a corporatized body.[70] Due to the low response rate and unavailability of data from public and corporatized water utilities, it was not possible to analyse the performance of this categories of companies. This question is revisited in Section 5.4 which contains a more detailed case study of PBAPP and SADA.

Private water sector companies achieved an average collection rate of 97% (Figure 5.3). As we only have data from one public sector company (SADA) it is not possible to draw a comparison between the two sectors. This also makes it difficult to derive a national average for collection efficiency for the benchmarking purposes. The question of collection efficiency and the effect that the reform may have had upon it are areas that warrant further research. Figure 5.3 does show that the private water sector companies have done reasonable well in reducing the percentage of uncollected bills, thus improving their revenue collection. Private water utilities are under pressure to achieve a healthy revenue stream. However, achieving an efficient collection efficiency rate requires investing in a new billing and collection system. For instance, after taking over water provisioning in the state of Selangor and the Federal Territories of Kuala Lumpur and Putrajaya in 2005, Syabas streamlined its billing and collection system by introducing an Integrated Water Management System (PNHB, 2009). Billing rationalization led to the collection efficiency to surpass 100% in 2006 and 2007. In 2008, it started to normalize. Its higher collection efficiency in 2009 was contributed by a 5.9% growth in active customer accounts (PNHB, 2009) and the lower collection efficiency in 2010 was affected by a 2.4% decrease in active accounts (PNHB, 2010).

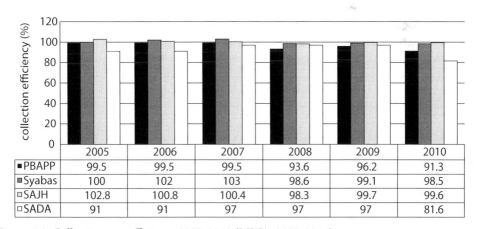

| | 2005 | 2006 | 2007 | 2008 | 2009 | 2010 |
|---|---|---|---|---|---|---|
| ■PBAPP | 99.5 | 99.5 | 99.5 | 93.6 | 96.2 | 91.3 |
| ■Syabas | 100 | 102 | 103 | 98.6 | 99.1 | 98.5 |
| ▫SAJH | 102.8 | 100.8 | 100.4 | 98.3 | 99.7 | 99.6 |
| ▫SADA | 91 | 91 | 97 | 97 | 97 | 81.6 |

*Figure 5.3. Collection rate efficiency 2005-2010 (MWA, 2005-2011).*

---

[70] SADA only became a corporatized entity from 1.1.2010. Previously water provisioning in the state of Kedah was done by JBA Kedah, the state water department.

In 2005-2007, a collection consolidation exercise led SAJH to register a collection efficiency level above 100%. This unusual high collection level occurred when collections from the previous year were brought forward to current year. However, its collection efficiency level normalized from 2008 (Ranhill Utilities Berhad, 2011).

It is not unexpected that private utilities had the financial capability to implement efficient revenue management. However, one might wonder how a newly corporatized water utility like SADA also managed to achieve good collection efficiency rates. This is discussed in Section 5.4.

*The connection between collection efficiency and revenue (2008-2010)*

It is not always true that a lower collection rate will cause a contraction in revenues. Despite recording lower collection efficiency, Syabas and SAJH still registered a growth in revenues in 2008 (Table 5.5). This growth was a result of an increase in active customer accounts for Syabas and 2.4% tariff increase in 2007 for SAJH (Ranhill Utilities Berhad, 2007). Syabas went on to record further revenue growth in 2009 and 2010. By contrast SAJH recorded a decline in revenue, despite registering higher collection efficiency in 2009-2010. Separate analyses for PBAPP and SADA are presented the case studies in Section 5.4.

In a wider context we can expect the reform to lead to further innovations in the (currently conventional) way bills are paid, especially among the recently corporatized water entities. Previously bill payments could only be made at post offices or at designated payment centres or agents. Now customers of SAMB, SAINS and SADA can pay their bills through the internet or using an auto-debit platform. All these innovations were introduced when business-minded management teams appointed to run the companies were given sufficient autonomy to implement measures to steer the companies towards profitability. A clear example was a measure implemented by SAINS to facilitate prompt bill payment. SAINS offered prizes – in cash and in kind – worth more than RM 100,000 to reward customers who paid at their payment counters. Figure 5.4 shows some of the advertising placed on their website for this campaign.

If SAINS' campaign to promote prompt payment of bills can be considered as providing a 'carrot', then consider Syabas's strategy of disconnecting water supplies for non-payment is a clear example of a 'whipping stick'. Syabas has been doing this on a regular basis and they disconnected

*Table 5.5. The effect of collection efficiency on revenue growth (2008-2010) (MWA, 2008-2011).*

| Year | Indicators | Syabas | SAJH |
|------|-----------|--------|------|
| 2008 | Collection efficiency (%) | 98.6 | 98.3 |
|      | Revenue growth (%) | 4.2 | 2.3 |
| 2009 | Collection efficiency (%) | 99.1 | 99.7 |
|      | Revenue growth (%) | 2.3 | -6.1 |
| 2010 | Collection efficiency (%) | 98.5 | 99.6 |
|      | Revenue growth (%) | 3.8 | -6.3 |

*Figure 5.4. SAINS' advertisement to encourage prompt bill payment by its customers (www.sainswater.com).*

7.8% more customers in 2009 than in 2008 (PNHB, 2009). Section 89 of the WSIA legalizes disconnecting a water supply for non-payment to act as a deterrent against (illegal) use of water (MEWC, 2006a). However, a disconnection can only be done when all other avenues have been exhausted. Alternatives must be sought beforehand. These could include pre-paid water meters (based on a mobile phone concept), which requires users to buy credit before they can use the water (Zaini, 2009; Berg & Mugisha, 2010). However this would require conventional meters to be replaced with ones that work with a pre-paid or special credit card (Van Vliet, 2002). It remains to be seen whether water utilities are willing to absorb the cost of installing new meters in exchange for higher collection rates.

### 5.3.3 Unit production cost: the art of managing the cost

(Public) water supply is considered as a 'service of general interest', vital for general welfare, public health and the collective security of people, as well as to economic activities and environmental conservation (Alegre *et al.*, 2006). Running a water supply system is a capital-intensive undertaking, no matter who is responsible for providing these services (MEWC, 2008). In most developing countries, the water tariff is not commensurate with the amount of investment made. Water Utility Management International (2009) reports only 31% of global water utilities operate in a full cost recovery environment. As such efficient cost management is a challenging for all water utilities. Usually water utilities that manage to keep their production costs below (or at par with) the prescribed tariff levels are on the right path for achieving profitability.

## Overall performance

It is interesting to note that there was a declining trend in the average production cost, even before the water sector was reformed (Figure 5.5). However, the average production cost did increase slightly in 2009.

## Performance by state water departments

As shown in Table 5.6, all the public water departments (except for JBA Kedah and JBANS) recorded a decrease in the average production costs in the period following the water sector reform.

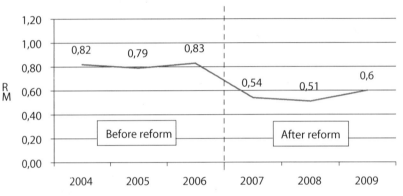

*Figure 5.5. The downward trend of overall production costs in RM/m³ (2004-2009) (MWA, 2004-2010).*

*Table 5.6. Average production costs of public water utilities before and after the reform (RM/m³) (MWA, 2005-2010).*

| State | Water utilities | Avg. production cost before reform (2004) | Avg. production cost after reform (2009) | % increase/decrease |
|---|---|---|---|---|
| Kedah | JBA Kedah | 0.36 | 0.48 | +37.5 |
| Labuan | JBA Labuan | 1.29 | 1.21 | -6.2 |
| Melaka | PAM/SAMB | 0.78 | 0.57 | -26.9 |
| N. Sembilan | JBANS/SAINS | 0.48 | 0.56 | +16.7 |
| Pahang | JBA Pahang | 0.58 | 0.47 | -18.9 |
| Perlis | PWD Perlis | 0.47 | 0.32 | -31.9 |

Despite decreasing their average production costs, some water utilities – particularly JBA Labuan had production costs that well above the national average. This, to a certain extent, reflects the inability of JBA Labuan to effectively manage its costs (despite an improvement). Nevertheless, in most cases, it is not yet possible to conclude that the reform was the only contributor to the decrease. Other factors may also have contributed to the decline in the average cost of producing one cubic metre of (treated) water.

Despite achieving a decrease in production costs, JBA Labuan faced a tremendous challenge in bringing down its production costs to a level which is at least at par with the prevailing tariff of RM 0.90/m$^3$ (MWA, 2010). A high usage of chemicals and energy consumption by its privatised Beaufort treatment plant contributed significantly to the JBA Labuan's overall increase in production costs. Large quantities of chemicals are needed to treat high turbid raw water from the Padas River, the main raw water source for the Beaufort treatment plant located in the mainland of Sabah. In addition, there is substantial energy costs involved in pumping treated water from this treatment plant through a submerged pipe to the island of Labuan. Beaufort plant supplies 80% of the total water demand on the island. As a result of these high production costs, JBA Labuan recorded a deficit of RM 5.9 million and RM 3 million in 2008 and 2009 respectively (MWA, 2010). If capital expenditure, depreciation and financial costs were to be included, JBA Labuan would record much higher overall production costs. However, as a federal water entity, these costs were absorbed by the federal government.

PAM/SAMB recorded the second highest drop in average production costs. Before the reform was introduced, one of its main costs was for pumping raw water from the Muar River in the state of Johor. This project – completed in 2003 – was implemented to avoid the occurrence of the 1991 water crisis caused by a prolonged drought and which affected more than 600,000 people (Angkasa Consulting Services, 2011). When SAMB came into existence as a corporatized entity in 2006, more prudent cost management measures for chemical and energy consumption were implemented in order to reduce the overall production costs. In 2009, SAMB managed to bring down its energy and chemical costs by 5% and 2% respectively (MWA, 2010). As a result of these prudent cost cutting measures, SAMB recorded a surplus of more than RM 6 million in the same year (MWA, 2010). This is in line with one of the aims of the reform: to steer public water utilities, such as SAMB, towards more efficient cost management.

JBA Kedah and JBANS recorded substantial increases in their average production costs, while PWD Perlis registered a drop. The increases for JBA Kedah and JBANS would have been higher if costs for capital expenditure, depreciation and capital cost were included. PWD Perlis would also most likely have registered an increase in production costs if these costs were taken into consideration. However, being the sole owner of JBA Kedah, JBANS and PWD Perlis, the state governments of Kedah, Negeri Sembilan and Perlis respectively absorbed these costs to avoid any unnecessary pressure on the water tariffs.

*Peformance by corporatized water entities*

Generally, except for SATU, the average production costs for corporatized water utilities decreased after the water sector reform (Table 5.7). SATU's increase in production cost – of a mere RM 0.02 – was not significant. All these entities were able to keep their average production cost below the

*Table 5.7. Average production costs for corporatized water entities before and after the reform (RM/ m³) (MWA, 2005-2010).*

| State | Water utilities | Avg. tariff[1] | Avg. production cost before reform (2004) | Avg. production cost after reform (2009) | % increase/ decrease |
|---|---|---|---|---|---|
| Kelantan | AKSB | 0.90 | 0.44 | 0.40 | -9.1 |
| Perak | LAP | 1.16 | 0.53 | 0.42 | -20.7 |
| Terengganu | SATU | 0.83 | 0.46 | 0.48 | +4.3 |

[1] Based on combined domestic and industrial 2009 tariff.

prevailing tariffs, and thus are on the right path for achieving profitability. AKSB and LAP both registered an increase in their revenue surplus in 2009, while SATU suffered a deficit as a result of lower revenues and increased operating expenditures (MWA, 2010).

At this juncture these changes in the average production costs cannot be explicitly linked to the reform. However, the decision to turn these utilities into corporatized water utilities – one of the main tenets of the reform – seems to have yielded favourable outcomes. Once they have stabilized and benefitted from the sustainable financing mechanism under the guardianship of PAAB (the national asset management company), one can expect them to be able to perform more efficiently, including in the area of cost management. According to the Director of the Water Supply Department at the MEWC[71], one of the options that water utilities could consider is to integrate small plants into big or regional plants which use less chemicals and have energy efficient equipment. Applying the latest technology, such as Geographical Information System (GIS) or Supervisory Control and Data Acquisition (SCADA) requires less manpower to operate the plants thus can reduce labour costs in long run.

Corporatization will also pave the way for the integration between water supply and the sewerage sector (into a single holistic water utility) which in the long run could bring down the production costs through a synergy in the management of human resources, equipment and facilities (Alegre *et al.*, 2006; Asian Development Bank, 2006). However, before this can happen, two fundamental issues facing the sewerage sector need to be addressed first: under-investment and tariff rationalization.

*Performance by private water entities*

Table 5.8 shows the tremendous decrease of the average production costs within the private sector after the reform was introduced. Syabas recorded the highest drop of 58% (see Section 5.4 for a case study of PBAPP).

Leaving aside the issue of financial costs, this table re-affirms that private sector companies are more efficient than the public sector in cost management. They are, to a certain extent, motivated

[71] Personal interview on 4 March 2011.

Table 5.8. Average production costs for private water entities before and after the reform ($RM/m^3$) (MWA, 2005-2010).

| State | Water utilities | Avg. production cost before reform (2004) | Avg. production cost after reform (2009) | % increase/decrease |
|---|---|---|---|---|
| Johor | SAJH | 0.88 | 0.62 | -29.5 |
| Penang | PBAPP | 0.43 | 0.40 | -6.9 |
| Selangor | Syabas | 1.23 | 0.52 | -57.7 |

by the desire to boost profitability and reducing production costs is one way of doing this. SAJH and Syabas managed to reduce their energy and chemical costs – the two main cost components of producing water – by more than the public sector companies (MWA, 2010). As a result, and the prevailing tariffs, they recorded surpluses of RM 133.9 million and RM 20.2 million respectively in 2009. Syabas's revenue was badly affected by the on-going dispute over the water tariff that it had with the state government.

Efficient cost management allowed private water utilities to accelerate cost recovery because of the financial obligation that was committed under the privatization scheme. Lowering the production cost meant that they can produce same of amount (treated) water with less investment, thus creating substantial cost savings for the company (Ranhill Utilities Berhad, 2009; Syabas, 2005). It also made it a little easier for both SAJH and Syabas to re-coup their huge investments, committed for the entire concession period, through the prevailing water tariff. Nevertheless, privatization without effective regulation, has led to the average end-users water tariff[72] for these two utilities to be among the highest in the country. This is in contrast to the claims that privatization results in lower tariffs (Holland, 2005).

Production cost is highly influenced by the level of NRW. A high NRW causes production costs to increase as less water is produced with the same amount of investment or more investment is needed to maintain quantity of water available. Both increase the final cost of producing one cubic metre of water. When NRW is included, the average production costs for the period after the reform for both SAJH and Syabas increased tremendously to RM 1.52 and RM 1.30 per cubic metre respectively (Table 5.9). This shows why water utilities around the world invest heavily to reduce NRW. It is not surprising at all to note that SAJH and Syabas are both committed to investing to the tune of RM 1.6 billion and RM 10.7 billion respectively, mainly for NRW reduction works.

### 5.3.4 Customer service complaints: the customer is king

According to AWWA (2004), the number of (service) complaints is a good indicator for measuring customer service. As in other service industries, a water utility can use the number of complaints to gauge customers' satisfaction with their services. Complaint management can provide a valuable

---

[72] RM 1.52 (Selangor), RM 1.92 (Johor).

Table 5.9. Effects of NRW on the average production costs after the reform (RM/m³)2007-2009 (MWA, 2008-2010).

| State | Water utilities | Avg. production cost after reform (A) | Avg. cost of NRW (B) | New avg. cost (A) + (B) |
|-------|-----------------|--------------------------------------|----------------------|--------------------------|
| Johor | SAJH | 0.62 | 0.90 | 1.52 |
| Penang | PBAPP | 0.40 | 0.49 | 0.89 |
| Selangor | Syabas | 0.52 | 0.78 | 1.30 |

tool to identify areas for improvement not only in the area of customer service, but perhaps also in other areas such as economic efficiency.

*Overall performance*

Figure 5.6 shows that the numbers of complaints per 1000 connections for all water utilities went down from 2007-2009 but increased again in 2010. Despite this increase the figure stayed equivalent to the average complaint level of 86 complaints per 1000 connections.[73]

Figure 5.7 shows a breakdown of the two categories of complaints, with 'other complaints' being the largest category. However, no further explanations on the elements that constitute 'other complaints' are given in the data. Even without an explanation, this is certainly of interest to the water sector. If the MWA report were to give detailed explanations of the nature of the complaints, this would be a great service to water utilities as it could help them formulate remedial actions, look at what urgently needs doing and what can afford to be delayed.

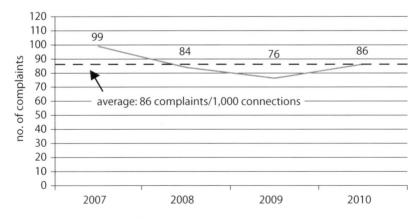

Figure 5.6. Overall number of complaints per 1000 connections (2007-2010) (MWA, 2008-2011).

---

[73] From 2007-2010.

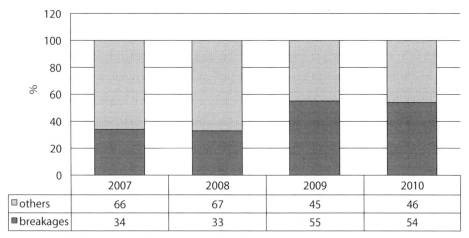

Figure 5.7. Breakdown on complaints category for all water utilities (2007-2010) (MWA, 2007-2011).

## Performance by ownership

In all categories of ownership, the number of complaints per 1000 connections demonstrated a decline except in 2010 when public water departments recorded a sharp increase and the corporatized ones a small increase. Complaints to private water utilities were stagnated in 2010 (Figure 5.8).

Overall, the numbers of complaints to most water utilities fall below the average of 86 complaints per 1000 connections (Table 5.10). However, this decrease cannot be taken as absolute indicator for measuring changes in the efficiency of the water utilities. For instance, even though both JBA Labuan and SAMB recorded the highest level of complaints, they fared better than many companies

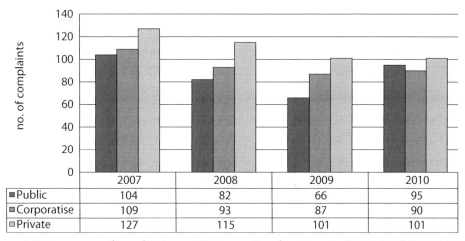

Figure 5.8. Average no. of complaints per 1000 connections by categories (2007-2010) (MWA, 2007-2011).

*Table 5.10. Average no. of complaints per 1000 connections (2007-2010) (MWA, 2008-2010).*

| Ownership | State | Water utilities | Avg. complaints per 1000 connections |
|---|---|---|---|
| Public | Kedah | JBA Kedah/SADA | 56 |
| | Labuan | JBA Labuan | 175 |
| | N. Sembilan | JBANS/SAINS | 101 |
| | Pahang | JBA Pahang | 57 |
| | Perlis | PWD Perlis | 93 |
| Corporatized | Melaka | SAMB | 144 |
| | Kelantan | AKSB | 93 |
| | Perak | LAP | 74 |
| | Terengganu | SATU | 66 |
| Private | Johor | SAJH | 76 |
| | Penang | PBAPP | 135 |
| | Selangor | Syabas | 122 |

in terms of NRW management. On the other hand, PBAPP which had the lowest NRW in the country recorded a much higher level of complaints than many utilities, such as JBA Pahang and AKSB which had a high level of NRW.

Some water utilities use complaints to not only measure the effectiveness of their service, but also as the platform for further improvements. Thus, some water utilities actively encourage customers to report on water problems. For instance, LAP provided incentive (10% of the amount of water bill) to water users who provided information about water theft (LAP, 2011). Furthermore, most of the water utilities have made it easier for customers to make complaints. Besides conventional methods (counter service), many are using information technology to create 'easy to use' complaints procedures for their customers. Customers now not only have access to 24-hours call centres (in the case of PBAPP and Syabas), but also can channel their complaints through websites and emails. A quick surf on the internet revealed that corporatized (SAINS, SADA, SAMB, and LAP) and private water entities were actively encouraging their customers to use on-line complaints forms in order to encourage water users to be more vigilant and communicative about water problems.

It can be envisaged that, when the remaining public water utilities are corporatized, they too will take advantage of information technology to improve their complaints procedures. Nevertheless, this matter goes beyond just accepting and handling complaints. What matters most is how water utilities react to those complaints, in terms of taking remedial actions and initiating follow-up so that the complaints are actually addressed.

## 5.4 Case study on operational efficiency: comparing private and public water utilities

### 5.4.1 Introduction

Section 5.3 presented an analysis about the relationship between the reform and the operational efficiency of water utilities. It examined, in a broad sense, the impact of the reform on water utilities from the perspective of ownership; comparing public (state department and corporatized) and private forms of organization. This section (Section 5.4), presents two case studies analysing the relationship between the reform and the operational efficiency of water utilities for (1) a private water entity PBAPP (Penang); and (2) a public water entity SADA (Kedah).

### 5.4.2 The impacts of the reform on PBAPP and SADA

The reform mostly had a direct impact on three aspects of the operations of PBAPP and SADA: (1) licensing and regulatory oversight; (2) tariffs; and (3) financing mechanism.

### Licensing and regulatory oversight

Under the new regime, it became mandatory for both PBAPP and SADA to obtain a license from NWSC, the central regulator, to continue their operations. In general, there are two types of licence: a 'service' licence for water treatment and distribution activities, and a 'facilities' license for owning water assets. Previously the state government issued licences to water operators. PBAPP and SADA both now come under the regulatory scrutiny of the NWSC, a function previously held by the state regulator. Penang Water Department regulated the behaviour of PBAPP, while before SADA was established the state of Kedah regulated its own water supply activities.

### Tariffs

Another major impact of the reform was on tariffs. The decision about the level of tariffs no longer rests with state governments but has been centralized under the NWSC. This mechanism is intended to eliminate political influence on tariffs at the state level (USAID, 2005). At the moment the water tariffs for PBAPP and SADA are both below the national average (see Appendix 4).

### Financing mechanism

Prior to the reform, PBAPP and JBA Kedah applied for interest free loans from the federal government to fund water infrastructure works and used tariff revenue to cover their operating expenditure. However, this financing arrangement was not sustainable, since it did not provide sufficient funds for water utilities to undertake water development works and the revenue generated through tariffs was insufficient to repay the federal loans. At the end of 2007, the debts outstanding to the state from all water companies amounted to about RM 7.6 billion (Hua, 2009). Hence, this arrangement was discontinued, and a national water assets company PAAB was established to assume the responsibility for securing the long-term loans required by water utilities (MEWC, 2008). PAAB now requires PBAPP and SADA to operate under 'the asset light model'[74], focusing on attaining greater efficiency in operations and maintenance as their core business activities (SPAN, 2011a).

### 5.4.3 Overview of the water supply sector in Penang

*Period before independence*

The history of water supply in Penang started back in 1804 when the island was part of the British Empire. Sir Francis Light, then British Governor of Penang, commissioned the development of the first simple water supply and distribution system to meet the needs of the 10,000 odd populations on Prince of Wales Island, as it was known at that time. In 1919, a water department within the George Town City Council was established headed by Mr. J.D. Fettes, the Municipal Water Engineer. The development of the water supply system continued until 1929 where Penang's first treatment plant in Air Itam commenced operations (PBAHB, 2009).

*Period after independence*

After Malaysia gained independence in 1957, water supply development, including in Penang, was further accelerated. In 1962, Penang's first dam, the Air Itam Dam was officially opened by the (then) Governor of Penang, His Excellency Tun Uda Al-Haj bin Raja Muhammad. On 1 January 1973, Penang's first water authority, Pihak Berkuasa Air Pulau Pinang (PBA), a state statutory body, was established. The role of the PBA was to supply treated water to the island and the mainland taking over the role of George Town City Council (serving the island) and the Public Works Department (of the federal government) which supplied water to the mainland (PBAHB, 2009).

In 1999, the PBA was corporatized into the PBAPP, a private limited company under the Companies Act 1965. Since then the PBAPP has been wholly owned by the State Government of Penang. The PBAPP was granted a license under Section 16 of the Water Supply Enactment 1988 to supply water for the whole of Penang. Selected fixed assets of the PBA including buildings, treatment plants and reservoirs were transferred to the PBAPP.[75] However, strategic fixed assets, such as dams, canals, catchment lands and recreational areas remained with the state government of Penang.

In 25 May 2000, PBA Holdings Bhd. (PBAHB) was incorporated as investment holding company to carry out all the business activities of the PBA Group of Companies.[76] Two year after

---

[74] Under the asset light model, PAAB will absorb the outstanding loans of the state government. In return, the state government will transfer their assets to PAAB at values to be negotiated and agreed.

[75] As required under Section 4 (2) of the Penang Water Authority (Successor Company) Enactment 1998.

[76] See Appendix 5 for list of companies under the PBAHB.

its inception, the PBAHB was listed on Malaysia Stock Exchange. In 2003, the PBAPP became the first water utilities to set up a 24-hour call centre in Malaysia. In the same year, it scored another first – receiving multi-site certifications for the treatment and the supply of water (under the ISO 9001:2000 series) from two international accreditation bodies – UKAS (UK) and DAR (Germany). In 2003, the PBAHB, through its associate company Pinang Water Limited, ventured into the Chinese water market securing 29-year concession rights to build and operate a treatment plant in Yichun City in Jiangxi Province. In line with business expansion, the PBAHB has diversified its business into water bottling, management consultancy and running a water supply training academy (PBAHB, 2009). PBAPP's water supply statistics for 2010 are shown in Box 5.1.

---

**Box 5.1. PBAPP water supply statistics (2010) (www.pba.com.my).**

| | |
|---|---|
| Area of Penang State: | 1,031 km$^2$ |
| Population (est.): | 1.61 million |
| Number of registered customers: | Total: 506,989 |
| | - Domestic: 439,728 |
| | - Trade: 67,261 |
| Number of customer care centres: | 9[a] |
| Water catchment area: | 62.9 km$^2$ |
| Number of dams: | 6 |
| Total raw water storage capacity: | 46,013 million litres |
| Main source of raw water: | Muda River |
| Number of treatment plants: | 10 |
| Designed capacity of treatment plants: | 1273.2 MLD |
| Number of treated water reservoirs: | 60 |
| Number of treated water towers: | 38 |
| Daily supply of treated water: | 957 million litres |
| Daily water consumption: | 782 million litres |
| NRW: | 18.2% |
| Total length of pipes (100 mm and above): | 3,981 km |

[a] In 2009 (PBAHB Annual Report, 2009).

---

### 5.4.4 Overview of the water supply sector in Kedah

The water supply in the state of Kedah is unique in the sense that it has element of both public and private operations. In fact it represents the whole range of basic models of water services described by Rouse (2007): municipal, corporatized-public, PPP and private.

*From state water department to state water corporation*

Before 2006, the provision of water supply in Kedah was the responsibility of the Public Works Department (PWD). The PWD is the federal agency and the technical arm of the Ministry of Works. Water supply is only part of the PWD's larger responsibilities.

In view of the rising demand for water, the state government decided to take over the role of PWD. In 2006, the state water department, JBA Kedah was established to handle most of the water supply in the state. Other portions of the business were contracted out to private operators, through various forms including service contracts and concession agreements.

When the WSIA came into effect in 2008 as the result of the reform, the state governments were asked to corporatize their water departments. JBA Kedah was corporatized on the 1st of October 2010 and was renamed SADA. SADA came under the supervision of the national regulator, the NWSC. A state water resources agency LUAN was established to oversee matters related to the management of water resources in the state.

*Private participation*

Private participation in the Kedah water supply provision takes two forms: service contracts and concession agreements.

*Service contract.* The first service contract between AUIB and the state government of Kedah was signed on 22 November 1990. Under this AUIB provides water to the mainland of Kedah (MEWC, 2006a). Two further agreements were signed in 1998 and 2000, bringing the total contract periods to 30 years ending in 2020. AUIB now manages five treatment plants and sells bulk water to SADA at an agreed rate.

*Concession agreement.* Under the Langkawi Water Privatization Agreement, signed in 1995, Taliworks Sdn. Bhd. (TWSB) was given a 25-year concession to supply water for the whole of Langkawi Island (MEWC, 2006a). Under the concession, TWSB was responsible for water treatment, distribution and issuing bills to customers. However, revenue collection remains with SADA. SADA's water supply statistics for 2009 are shown in Box 5.2.

### 5.4.5 Explaining the relationship between water sector reform and operational efficiency

In this section I first examine the performance of both PBAPP and SADA with regard to operational efficiency. In the second part, I analyse how PBAPP and SADA reacted to the reform and what changes have taken place within the two water utilities as result of the reform.

*Non-revenue water*

The overall NRW level for both utilities seems to be quite stable, as shown in Figure 5.9. PBAPP's NRW levels declined before the water sector was reformed, and continued to decline further after

---

**Box 5.2. SADA water supply statistics (2009).**

| | |
|---|---|
| Area of Kedah State: | 9,500 km$^2$ [a] |
| Population (est.): | 1.94 million[a] |
| Number of registered customers: | Total: 506,068[b] |
| | - Domestic: 457,957 |
| | - Trade: 48,111 |
| Number of customer care centres: | 1[b] |
| Water catchment area: | Data not available |
| Number of dams: | 4[b] |
| Total raw water storage capacity: | Data not available |
| Main source of raw water: | multiple sources |
| Number of treatment plants: | 33 (2010)[b] |
| Designed capacity of treatment plants: | 1,612[c] |
| Number of treated water reservoirs: | Data not available |
| Number of treated water towers: | Data not available |
| Daily supply of treated water: | Data not available |
| Daily water consumption: | 646 million litres[c] |
| NRW: | 44.9%[c] |
| Total length of pipes: | 14,644 km[c] |

---

[a] The National Statistics Department, 2011.
[b] As of 2010 (exclude privately-owned plants) – data obtained from questionnaires.
[c] MWA, 2010.

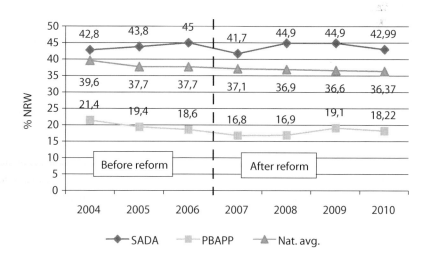

*Figure 5.9. NRW PBAPP, SADA and the national average (2004-2010) (MWA, 2005-2011).*

the reform except in 2009. Conversely, the situation at SADA showed the opposite trend, except for 2007. However, both appeared to be managing their losses better in 2010. SADA's NRW level is above the national average, and there are pressures on the company to reduce its NRW to the 20% target threshold proposed under the reform, (MEWC, 2008).

*PBAPP*

Industry practitioners regard PBAPP's achievement in maintaining the level of NRW below 20% for many years as highly commendable. Other private entities, such as Syabas and SAJH do not even come close to what PBAPP has achieved. PBAPP performs satisfactorily compared with other water utilities in the region (Figure 5.10). Yet these credible results may have led PBAPP to feel a little complacent and not pay attention to the possibility that the natural rate of rise in leakages might have an effect. NRW rose by 2.2% in 2009 (MWA, 2010). When asked to identify the root causes for the increase, PBAPP's Strategic Planning Manager stated that 'since PBAPP has achieved 'good' NRW, the management decided to trim the budget for NRW works'.[77] This goes to highlight the importance of continuously monitoring NRW reduction programmes.

In 2010, PBAPP decided to use its remaining allocation under the 9th Malaysia Plan to address NRW and used almost 100% of the RM 50 million allocated for NRW works (Economic Planning Unit, 2008). This led to the NRW level dropping to 18.2%.[78] For the General Manager, the challenge

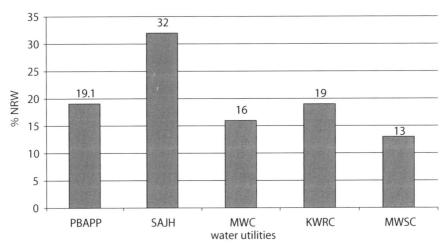

*Figure 5.10. NRW of PBAPP and selected regional water utilities (2009).*
*Note: PBAPP – Perbadanan Bekalan Air Pulau Pinang[1]; SAJH – Syarikat Air Johor Berhad, Malaysia[1];*
*MWC – Manila Water Company, the Philippines[2]; KWRC – Korea Water Resources Corporation,*
*South Korea[2]; MWSC – Macao Water Supply Company, Macao SAR[2].*
*Sources are [1] MWA, 2010 and [2] WaterLinks Forum, 28-30 Sept. 2009, Bangkok, Thailand.*

---

[77] Personal interview on 12 Nov. 2009.

[78] Personal interview with the Strategic Planning Manager on 12 Nov. 2009.

for PBAPP is to sustain the level of NRW achieved thus far. He is confident that PBAPP can meet the 18% and 16% targets for 2011 and 2012 that it has promised NWSC (PBAPP, 2011). Furthermore, the 2011 National Water Services Master Plan Study (jointly conducted by MMC and Sumitomo Corporation) quoted that PBAPP has achieved 'break even'[79] on NRW (MMC-Sumitomo, 2011). This raises questions as to whether it is economically viable for it to seek to further reduce NRW, as this would cost more money than the benefits it generates (MMC-Sumitomo, 2011). An alternative strategy for PBAPP could be to channel those investments into water production, to offset the amount of water loss (rather than further reducing the NRW).

The reform seems to have little impact on PBAPP's NRW efficiency, which was already almost twice the national average. Indeed in a more general sense we cannot expect such a policy intervention to be likely to influence the performance of water utilities in this area in such a short period of time. Even before the reform, PBAPP managed to sustain its NRW levels well below 20% threshold. In fact, its commendable performance was quoted being as the economic level of leakages for the whole country (MEWC, 2008). Thus, it is likely that its performance was more due to on-going NRW reduction activities implemented prior to the reform, a view confirmed in interviews with several top managers of the company. According to PBAPP's General Manager, Jaseni Maidinsa, their framework, looks at addressing losses right from where water is treated (treatment plant), through the distribution systems and to where it passed onto the end consumer (at the metre). PBAPP was among the first water utilities (the other is LAP) to set up a dedicated NRW Team. According to PBAPP's Strategic Planning Manager, the NRW Team enabled PBAPP to focus its resources 'to sustain the control mechanism already in-place and to drive continuous improvement'.[80] Its General Manager believed the company deserved more recognition for this: 'since PBAPP is acknowledged and accepted as the best managed water utility in the country (in terms of NRW and other areas)', but puzzled as to 'why the (federal) government is not willing to adopt our model'.[81]

Data obtained from the questionnaire survey reveals that almost 80% of the leakages suffered by PBAPP consist of real or physical losses. The other 20% comes from apparent or commercial losses. Thus, undertaking programmes to combat real losses on a continuous basis remains the main priority of PBAPP. To this end, PBAPP's quality improvement commitments that involve strict pipe replacement procedures, speedy and quality pipe repairs, effective pressure management, active leakage control and District Metering Areas have contributed significantly to reducing NRW levels. In a paper presented in Malaysia Water, Jaseni Maidinsa (2009) states that if real losses can be minimized, more water can be sold to generate revenue. PBAPP's Strategic Planning Manager added that reducing real losses has a much bigger impact 'as water recovered through leakage is in fact a resource, reducing real losses means less money will be spent on developing new resources'.[82] He also recognizes that any additional water resources for PBAPP would have to come from outside the state of Penang, and this would be a costly affair. From an environmental

[79] Refers to a situation where costs to replace pipes is higher than costs of water losses.
[80] Personal interview on 12 Nov. 2009.
[81] Personal interview on 4 Nov. 2009.
[82] Personal interview on 12 Nov. 2009.

perspective, reducing water losses could delay (or eventually cancel) the development of new water resources; hence minimizing environment impacts.

## SADA

In contrast to PBAPP, reducing NRW levels is one of the great challenges for SADA. Figure 5.9 shows that the NRW levels for SADA have been as high as 45% (in 2006). And, even after a decrease to 41.7% in 2007 they returned to 44.9% in 2008 and 2009 before declining again slightly in 2010. Such high levels clearly must have an impact on SADA's financial standing. With almost half of the water that could be sold, lost, the financial losses were huge. In 2008 and 2009 the water losses amounted to 190,581 m$^3$ and 192,586 m$^3$ respectively (MWA, 2010). At the average tariff of RM 0.87 per m$^3$ (in 2009), the cumulative financial losses must have been in excess of RM 333,355. Such high losses clearly are an obstacle to SADA becoming financially viable. These figures clearly show that SADA's NRW reduction programmes in recent years have not been effective.

Before questioning the effectiveness of the reform in improving the efficiency of SADA, one has to understand the history of water provisioning in the state of Kedah, which has been operated as a public service for a long time. From independence until 2006, water supply was handled by the PWD. And as in other developing countries, the main objective of PWD at that time was not on water efficiency but on promoting public health, through providing clean water (Kibassa, 2011). The PWD also handled other portfolios including roads and bridges. Having to compete with other portfolios for limited budgets, the water sector was often neglected.

The issue of inadequate funding continued even when water provisioning was taken over by the JBA Kedah in 2006. The JBA Kedah relied heavily on federal budgets for capital expenditure works. Under the 9$^{th}$ Malaysia Plan, JBA Kedah was given RM 100 million for its NRW reduction programmes, but only managed to spend RM 53 million or 53% of this approved budget (Economic Planning Unit, 2008). If the JBA Kedah had been able to use the budget more efficiently, they could have reduced many of the existing 5,816 km of leak-prone asbestos cement pipes and replaced many of the 219,722 old water meters (aged above 7 years) (MWA, 2010). Unfortunately, there was no regulating body to oversee JBA Kedah, or impose penalties for missed targets. This provides clear evidence of a severe conflicts of interests, since JBA Kedah functioned both as a service provider and a regulator. This may well explain the establishment of SADA in 2010. While there are historical reasons why JBA Kedah experienced a high NRW, the following paragraph looks at the likely impacts of the reform on NRW, whether through changes to the role of the state government or the new role of SADA.

The reform was perhaps the turning point, which pressured the state government to corporatize JBA Kedah. Without the reform, the decision to corporatize JBA Kedah might have taken a longer to materialize. However, it is too early in the day to make any evaluation of SADA's performance within such a short span of time. Corporatization does indicate the (political) will of the state government[83] to improve the efficiency of water supply for its citizens and promote good governance (Ehrhardt & Janson, 2010). Nevertheless, the current NRW level is clearly not tenable for a 'business entity' and SADA is not able to generate sufficient revenue to sustain its

---

[83] The state government of Kedah is governed by an opposition party since 2008.

operations effectively. According to SADA's General Manager (Technical), further water losses have to be contained, and this will involve SADA 'in implementing District Metering Zones and a Geographical Information System in stages'.[84] He is confident that the implementation of these and other plans will enable SADA to meet the NRW targets it has promised the NWSC: 42% in 2011, 40% 2012 and 38% 2013. As a corporatized water entity SADA will be subject to tight regulatory oversights from the NWSC but also benefit to access to financial resources from the PAAB. He believed that these arrangements will allow SADA to manage its NRW reduction works effectively and that the 'plan to reduce NRW to 30% by 2020 is reachable'.[85]

Governments that have initiated public sector reform are often asked how corporatized state water companies can improve NRW management (Bhuiyan & Amagoh, 2011). On a theoretical level, corporatization allows a public utility to have their operations more separate from the command and control of the state (Rouse, 2007; Fuest & Haffer, 2007). In this regard, it seems that SADA is being allowed to implement the measures it views necessary to promote efficiency, without unnecessary interference from the state. Those interviewed confirmed this view. The first confirmation came from the highest decision-making body in the federal state's administration. In that interview, the State Minister responsible for water supply said that 'state government is giving SADA a free hand to determine its own course. We do not meddle with their operation'.[86] The willingness of the state government to relax control over SADA is highly welcomed at the working level. SADA's Customer Service Manager also acknowledged that, in terms of day-to-day operations, 'the Board gives our CEO a free hand in managing the company'.[87] Having autonomy, especially the right to spend the revenue it generates, allows SADA to formulate comprehensive NRW reduction programmes, commensurate with the financial resources at its disposal. This in contrast to the previous situation where water revenue was consolidated, as part of the state's income and went into state coffers.

## Collection efficiency

Overall, both PBAPP and JBA Kedah/SADA have achieved satisfactory levels of collection efficiency: above 95% (Figure 5.11). PBAPP's collection efficiency level remained relatively stable between 2005-2008 but declined somewhat in the following two years. JBA Kedah/SADA's efficiency level stabilized at 93% from 2007-2009 but also declined (quite substantially) in 2010. It is interesting to note that JBA Kedah/SADA has managed to increase its collection efficiency since 2007 (Figure 5.11). This goes to show that it is possible for a public water utility to increase its efficiency of collection.

---

[84] Personal interview on 6 Aug. 2010.

[85] Personal interview on 6 Aug 2010.

[86] Personal interview on 19 July 2010.

[87] Personal interview on 9 March 2011.

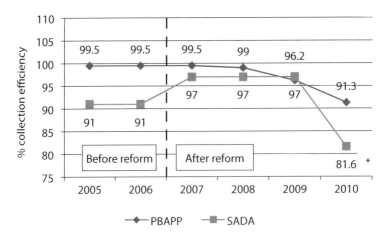

*Figure 5.11. Collection efficiency rate (2005-2010) (MWA, 2005-2011).*
*\* as of July 2010.*

*PBAPP*

Despite the drop recorded in 2009 and 2010, PBAPP generally managed to register a high average collection efficiency of 97%, a level which deserves a special mention. However, it would be a mistake at this point in time to link any change in the level of collection efficiency to the reform. The reform which was only implemented in 2007 is unlikely to have any significant effects on this performance indicator. It is more likely that the increase in average collection efficiency to 97% was caused by other factors. The analysis of available information shows two likely contributory factors: the introduction by the company of an Integrated Revenue Management System (IRMS), and a high willingness to pay among its consumers.

*IRMS.* PBAPP recognizes the need to have steady income streams and proper cash flow management. Back in 1975, its predecessor, the Penang Water Supply Authority, was the first company in Malaysia to introduce a computerized water billing system. PBAPP has continued to innovate. In 2002, a fully integrated billing system – IRMS – was implemented (MWA, 2008b). This provides on-line links between IT operations, corporate services, operations and finance databases (Jaseni, 2009). IRMS has contributed to reducing the number of (manual) billing errors. The use of IT platforms enabled PBAPP to link its payment operations with several participating agents, such as post offices and commercial banks. As well as making payment at PBAPP's customer service counters, water users in Penang can pay their water bill through automated teller machines (ATMs), by credit cards or bank drafts, or through 24x7 internet banking (PBAHB, 2010).

*Willingness to pay.* PBAPP regards customer satisfaction and continuous improvement as vital for its business survival (PBAHB, 2009). Like other service industries, PBAPP sees that customers who are willing to pay (the bill) help revenue generation. Customers will not pay if they are not satisfied and lack confidence in PBAPP's ability to deliver good water and good services. In this respect, high collection efficiency is a result of its customers' high willingness to pay (bills) and is a testimonial to PBAPP's good track record in the water supply business. It reflects customers' overall confidence in PBAPP as an efficient water utility. For efficient billing management, PBAPP embarks on various continuous improvements which include integrating both human resources and financial databases under a single platform. This seamless operation allows the PBAPP to store its billing water system, GIS and SCADA systems within one cohesive and secure enterprise data management system. In its official website, PBAPP has announced that for 2011 and 2012, its collection target is 96% (PBAPP, 2011).

## SADA

Before 2006, water provisioning in Kedah was under the PWD. In 2006, this responsibility was handed over to JBA Kedah, and for the first time a dedicated department was created to handle water business in the state. The establishment of JBA Kedah was in anticipation of the policy call from the federal government to corporatize state water departments as part of the reform. The (then) Director of JBA Kedah[88] stated that he got full political support from the state government in implementing the road map for this change, which included enhancing the efficiency of the existing bill collection system. In addition to conventional payment modes (e.g. post offices), JBA Kedah widened its bill payment platform to include 24-hour on-line bill payment[89] in collaboration with several major banking groups such the Maybank Group and CIMB Group. This arguably contributed to the increase in collection efficiency to 97% in 2007-2009. For a state water department, such as JBA Kedah, to get above 95% collection efficiency level is a great achievement. It has set a (high) benchmark for other state water departments to follow.

It is envisaged that, to certain extent, the corporatization of JBA Kedah could encourage further improvement in collection efficiency especially when the existing IT platform is upgraded under the company's ICT Blueprint, (which would make it similar to PBAPP's IRMS). SADA believes that the implementation of the ICT Blueprint will allow it to maintain the momentum achieved to date and is confident in meeting the collection efficiency targets pledged to NWSC: to achieve 98% and 98.5% collection efficiency rates for the years 2011 and 2012 respectively. Its achievement of these targets will be closely scrutinized by both the NWSC and the public, particularly the Water Forum, the public watchdog established under Section 69 of the WSIA (MEWC, 2006a).

## Collection efficiency and revenue

Even though collection efficiency is not the only variable affecting the revenue of particular water utility, it does, to a certain extent, determine the profitability of water utilities. It is generally believed, though not always true, that a water utility with lower collection efficiency rate tends to record lower revenues and thus profits. This was the case with PBAPP and SADA.

A decline in collection efficiency (of 2%) in 2009 contributed to PBAPP's revenue being 3.3% down on the previous year's figure (from RM 200.9 million in 2008 to RM 194.5 million in 2009 (MWA, 2009). Yet there was also a 1.0% (or 2,676,547 m$^3$) reduction in overall water consumption in the same year, another contributory factor (PBAHB, 2009). This said, the lower collection

---

[88] Personal interview on 11 May 2009.

[89] It is now common in Malaysia that water bills can be paid via internet banking.

efficiency did not severely affect PBAPP's profitability. PBAPP still managed to record (reduced) after tax profits of RM 14.8 million in 2009 (PBAHB, 2009), and a 77% increase in after tax profit (to RM 26.3 million) in 2010 (PBAHB, 2010). These were achieved despite the company having the cheapest water tariffs in the country (see Appendix 4). PBAPP's revenue stream was kept healthy because it has the lowest operating and maintenance cost per connection, by a 41.4% revenue growth (MWA, 2011), by achieving 18.2% NRW level, a 9.3% increase in trade water accounts, and a 5% drop in administration expenses (PBAHB, 2010). PBAPP is expected to record revenue level from the new tariff structure enforced in November 2010 which for the first time will impose a water conservation surcharge for consumption above 35,000 litres per month.

SADA recorded an increase in revenue from RM 172.3 million in 2007 to RM 175.0 million in 2008 (and to RM 180.7 million in 2009) as a result of higher collection efficiency (MWA, 2009, 2010). In 2010, its revenue stood at RM 214 million (MWA, 2011). However, due to higher operating and maintenance costs per connection, SADA recorded a deficit of RM 79.3 million in 2007. SADA returned to the black in 2008 and 2009 with surpluses of RM 3.9 million and RM 8.0 million, respectively. The surplus was quite small and can only cover operating expenditure with SADA having to depend on loans from the federal government for capital expenditure. The surplus could have been higher if SADA had managed to reduce its NRW. In 2010 when it had a lower NRW rate of 42%, SADA managed to record a higher surplus of RM 23.1 million (MWA, 2011). Moreover, for SADA the current tariffs, in place since 1993, have fallen way behind the increase in production costs. On average SADA's tariffs are among the lowest in the country (see Appendix 6) and these are being reviewed by the state government.

Two observations can be made about tariff rationalization. The tariffs of both PBAPP and SADA were rationalized after the reform and they were implemented when both the states of Penang and Kedah were under the opposition control.[90] For the record, the tariffs for both PBAPP and SADA had not previously been reviewed since 2001 and 1993 respectively (MWA, 2010). Tariff rationalization shows the growing awareness from, and willingness of the state governments to recognize the full cost of water, and to gradually reduce subsidies on water tariffs. Both administrations recognized the current tariffs were an impediment to achieving long-term sustainability in the water sector. Perhaps the most welcoming indication was that the (opposition) government in both states showed they willing to 'take politics out of water' (depoliticize water).

Chan (2009) sees the attempt to minimize (or possibly eliminate) political interference as a prerequisite for promoting more effective water governance in Malaysia. This was made possible by having the task of policy formulation, regulation and service delivery separated (Nyarko, 2007). As an economic (and also technical) regulator, NWSC, to a certain extent, removes the 'political liability' of state governments, thus minimizing political influence in tariff setting. Perhaps this will augur well with promoting good governance in the water sector, where decisions over water tariffs should be based solely on socio-economic reasons rather than on political considerations (as has been the case until today).

---

[90] The state of Penang was governed by a coalition government led by the Democratic Action Party, while Parti Islam SeMalaysia led the coalition government in the state of Kedah.

*Unit production cost*

As shown in Figure 5.12, PBAPP recorded a decrease in its average production costs after the water sector was reformed while SADA recorded an increase. However, the average production costs for both companies stayed well below the national average after the reform – even though this declined dramatically.

Beside basic costs for the treatment and distribution of water, another main cost component is the financial cost. In most cases, the financial costs for public water departments are absorbed by the state government. By contrast, private water operators are responsible for their financial costs, which are normally recouped through water tariffs. Table 5.11 which shows these figures after the financial cost is added, shows that PBAPP recorded an increase of 28% in the average production costs after reform, (still far lower than the national average) while for SADA the total cost of producing one cubic meter of treated water remained unchanged.

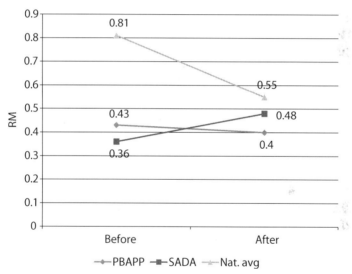

Figure 5.12. The average production costs before and after the reform (RM/m³) (MWA, 2005-2010).

Table 5.11. Effects of financial costs on overall production cost (RM/m³) (MWA, 2008-2010).

| Water utilities | Avg. production cost after reform | Avg. production cost after reform (financial cost added) | % increased/ decreased |
|---|---|---|---|
| PBAPP | 0.40 | 0.51 | 28 |
| SADA | 0.48 | 0.48 | 0 |
| Malaysia | 0.55 | 0.87 | 58 |

*PBAPP*

Table 5.11 shows how financial costs affected the average production costs for PBAPP. Despite an increase of 28% in production cost, PBAPP managed to sustain its overage cost below the national average. Unlike SADA, PBAPP is obliged to use a portion of its revenue to repay loans to the federal government, one component of its financial costs. The other components are capital expenditure and depreciation. The overall increase in production costs has affected the profitability of PBAPP. It's after tax profits plunged from RM 51.6 million in 2007 to RM 27.8 million in 2008 and fell again in 2009 to RM 15.8 million. The decreased profit recorded in 2009 was also associated with an increase in administration costs of RM 2.7 million that was incurred as a result of a 3.6% growth of registered water consumers and programmed improvements made to customer services (PBAHB, 2009).

PBAPP still manages to record a decent profit, despite the rising production costs and a low water tariff. Nearly 87% or 439,728 of its domestic customers pay the lowest tariff in the country, RM 0.31/$m^3$, while the other 13% of trade consumers pay RM 0.94/$m^3$. The current domestic rates do not reflect cost recovery. In the long run, this tariff is not sustainable and it is estimated that at current rates,[91] PBAPP subsidizes up to 60 $m^3$ of water per month per household. PBAPP's General Manager was convinced 'that a tariff increase is inevitable for PBAPP to continue serving their customers'.[92] This process of rationalization began in November 2010. The trade tariff was raised by 27% across all categories and for the first time, a water conservation surcharge was imposed for domestic consumption (see Box 5.3). The trade tariffs mainly affect large-scale water users such as industry, manufacturing, shipping and commercial enterprises. Despite the increase, the trade water tariff in Penang is still amongst the lowest in Malaysia, and in the Asia region (PBAPP, 2011).

*SADA*

Financial costs did not affect the average production costs for JBA Kedah as the state government absorbed the cost for loans repaid to the federal government (Table 5.11). Thus, it was no surprise to learn that SADA recorded a higher surplus for the 2009 financial year, despite the increase in production costs required to meet consumption growth of 3.3% (MWA, 2010). SADA would have been able to achieve higher profits if it had been able to reduce its NRW level.

When SADA can prove that it can stand on its own feet and make a decent profit, the state government might want it to claim the cost for loan repayments. When this happens, the company's overall costs will certainly go up. As a corporatized entity SADA must take some responsibility for its financial matters. In the long run, this could turn out to be beneficial in inculcating a culture of good financial management within SADA to sustain its place in the water sector where profits and earnings are highly regulated.

The reform has motivated SADA to become more concerned about efficient cost management, which will determine the profitability of SADA as a business entity. Energy is one of the major cost components for SADA and represents almost 70% of the overall cost of producing one cubic

---

[91] Enforced since 2001.
[92] Personal interview on 4 Nov. 2009.

---

**Box 5.3. PBAPP new tariff structure (as of 1.11.2010) (PBAPP leaflet, 2009).**

**Domestic tariff**

| | |
|---|---|
| Minimum charges: | RM 2.50 per month |
| First 20,000 litres: | RM 0.22 per 1000 litres |
| 20,000 litres to 40,000 litres: | RM 0.42 per 1000 litres |
| 40,000 litres to 60,000 litres: | RM 0.52 per 1000 litres |
| 60,000 litres to 200,000 litres: | RM 0.90 per 1000 litres |
| More than 200,000 litres: | RM 1.00 per 1000 litres |
| Water conservation surcharge | |
| Consumption above 35,000 litres per month: | RM 0.24 per 1000 litres |

**Trade tariff**

| | |
|---|---|
| Minimum charges: | RM 10.00 per month |
| *Trade ordinary* | |
| First 20,000 litres: | RM 0.66 per 1000 litres |
| 20,000 litres to 40,000 litres: | RM 0.89 per 1000 litres |
| 40,000 litres to 200,000 litres: | RM 1.15 per 1000 litres |
| More than 200,000 litres: | RM1.27 per 1000 litres |
| *Trade (special)* | |
| Flat rate: | RM 1.52 per 1000 litres |
| *Trade (shipping)* | |
| Flat rate: | RM 2.54 per 1000 litres |

---

metre of treated water. This had prompted SADA to launch an energy audit to seek to reduce energy consumption at all its treatment plants. A dedicated team, led by a senior engineer, has been formed to take on this task. He envisaged that the company would demonstrate much greater commitment in ensuring efficient energy usage in its treatment plants, and to install energy efficient equipment in the future.[93] However, the company's immediate response has been to request a favourable energy tariff from TNB, the national energy company as SADA currently pays the normal energy tariff, despite being a large customer.

## *The impact of the reform on production costs*

Water utilities believed that after the reform they would incur higher costs to meet the regulatory requirements of the NWSC. Many water utilities, especially in the public domain, are not ready to meet most of these regulatory requirements. They would need, for example, to make huge investments in pipe replacement to meet the NRW target, in IT for databases, in human resources, and in meeting environmental regulations.

---

[93] Personal interview on 19 July 2010.

SADA applied for RM 233 million under the 10th Malaysia Plan for water infrastructure works, of which RM 63 million was for NRW reduction programmes. PBAPP applied for RM 317 million for the same period, of which RM 20 million was for NRW related works (Economic Planning Unit, 2010).

There are also costs in meeting environmental regulations for sludge management. At the moment, only 35% of SADA's treatment plants[94] comply with the Environmental Quality Act (EQA) 1974. Only 30% of PBAPP 10 treatment plants meet the EQA 1974 requirement of having on-site sludge treatment facilities. If both SADA and PBAPP were to migrate to green technology for all plants, there is no doubt that their overall production costs would increase significantly. Both SADA and PBAPP argue that if they are required to migrate to environmental-friendly technology, they must be allowed to recover the extra cost through higher tariffs. However, according to PBAPP's Production Manager,[95] imposing such a 'green tariff' might not be a good idea at this point in time. He does not think that consumers would appreciate the benefits of such a move yet.

*Customer service complaints*

Figure 5.13 showed that the number of complaints per 1000 connections for both PBAPP and SADA dropped between 2007 and 2009. Figure 5.13 also disproves the assumption that water utilities, which manage higher number of connections record proportionately higher numbers of complaints. In 2009 for instance despite managing higher numbers of connections than PBAPP, SADA recorded less complaints per 1000 connections (MWA, 2010).

AWWA (2004) has suggested that customer complaints are a good measure of the efficiency of water utilities. However, not all instances of customer complaints are a good indicator of operational inefficiency. The level of complaints is not an accurate measure of the level of efficiency. For example SADA recorded much lower numbers of complaints per 1000 connections than the

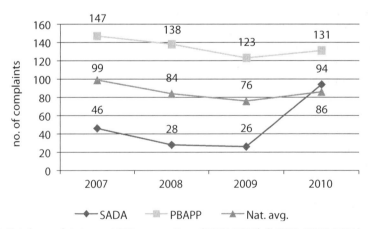

*Figure 5.13. Total complaints per 1000 connections (2007-2010) (MWA 2007-2011).*

---

[94] Not included treatment plants managed by AUIB and Taliworks.

[95] Personal interview on 2 Nov. 2009.

PBAPP. Can we say that SADA is more efficient than PBAPP? The answer is no. Statistics from the MWA showed that almost in all aspects of operational efficiency like NRW, collection efficiency, etc.; SADA fared less favourably than the PBAPP.

Moreover, some water utilities, such as PBAPP, have invested millions in user-friendly complaints technologies (such as web-based ones), actively inviting complaints – not only to gauge their level of services, but also as a tool for further improvement. In this situation, it is logical to expect that PBAPP will receive a higher number of complaints than public water utilities (like JBA Kedah/SADA) which still depend on conventional methods, or are at the initial stage of deploying complaint management technology.

## PBAPP

PBAPP views customer complaints as providing an opportunity for it to make improvements and, for this reason, complaints are taken seriously. For example, in 2008, PBAPP handled 1,421 complaints about burst pipe: 97% of which were repaired within 24 hours of being reported (PBAHB, 2008). In 2009, the repair rate within 24 hours went up to 99.6%. In the same year, 5,513 km of pipes were scoured and 26 km of old and leakage-prone asbestos cement pipes were replaced (PBAHB, 2009). This has contributed significantly to the continuous supply of good quality water to their customers in Penang.

PBAPP was among the first water utilities to establish an online one-stop customer care centre in 2001. A year later, PBAPP launched its 24-hour call centre. In the following year, a dedicated customer email service – customer@pba.com.my – and a corporate website – www.pba.com. my – were rolled out to facilitate smoother operations and efficient billing collection (PBAHB, 2008). In 2009, the customer care campaign – with the tag line 'Friendly, Caring, Responsive' – was launched. This tag line has since become the mantra of the company's management of customer complaints (PBAHB, 2008).

Its call centre allows the PBAPP to keep close contact with its customers, and effectively attend to their problems. It is the call centre that receives and addresses most of the complaints. In 2008, PBAPP managed to resolve 75% of the complaints received during the first contact with the call centre (PBAHB, 2008). According to Customer Call Centre,[96] located in Perai, the two most common complaints were about burst pipes and interruptions to the availability of water supply.

PBAPP aspires to continuously improve every aspect of its operations. Two areas that they are currently working on include consumer engagement and installing new technology for handling complaints. A Customer Service Manager[97] is responsible for these tasks and is planning to establish a customer engagement team to proactively address customer complaints. The manager stressed the priority that PBAPP gave to lifting improving its customer complaints management. For this end, she said that PBAPP used benchmarking and compares itself with two other utilities (SAJH and Syabas), which both also serve urban populations.

Table 5.12 shows the customer complaints targets that PBAPP has set itself (and pledged to the NWSC to keep) for 2011 and 2012.

---

[96] Personal interview on 3 Nov. 2009.
[97] Personal interview on 26 Oct. 2009.

*Table 5.12. PBAPP's complaint targets for 2011 and 2012.*

| Type of complaints | Targets baseline (2008) | 2011[a] | 2012[b] |
|---|---|---|---|
| Billing complaints: % responded to within 3 working days | NA | 85 | 95 |
| Response to complaints: % complaints meaningfully responded to within 5 working days | 80 | 85 | 95 |
| Telephone complaints: % responded to within 30 seconds | 100 | 100 | 100 |

[a] www.pba.com.my.
[b] PBAPP documents.

## SADA

When water provisioning was managed by PWD, complaints were manually recorded and there is a chance that some complaints were not recorded or are missing from the database. When JBA Kedah assumed the role of water provider in 2006 it consolidated the data it had – and there is a possibility that more data was lost. Both these factors may have contributed towards the low level of complaints recorded by JBA Kedah from 2007-2009 (MWA, 2009, 2010).

The first attempt to set-up a centralized complaint management was made in October 2009. An Information Centre was established to handle complaint management for JBA Kedah, and later in anticipation of the corporatization of JBA Kedah. When SADA came into existence in 2010, complaints management was centralized at its headquarters in the city of Alor Setar. However, SADA's Information Centre is far less accessible than PBAPP's Customer Call Centre only being open for 16 hours a day.[98] Appendix 7 explains how SADA's complaint management works.

SADA acknowledges the importance of forging active engagement with its customers. A community engagement programme called Rakan SADA or Friends of SADA was launched on 11 May 2011 by the Chief Minister of Kedah. To launch this, 1000 rural community leaders were appointed to assist SADA in tackling water problems (Warta Darulaman, 2011). SADA's Customer Service Manager said that 'Rakan SADA is a clear example of SADA's continuous commitment towards providing best services to our customers'.[99]

SADA has created a web-based platform to facilitate systematic and effective complaint management. Now water users in Kedah can register complaints through a toll free number, or use an e-complaint[100] platform on the company's website, www.sada.com.my. Reliable and accurate information enables SADA to formulate action plans to address water-related issues raised by the customers. SADA's complaints management system could come under tighter scrutiny from the NWSC. Section 15 of the SPANA 2006 (MEWC, 2006b), states that the NWSC should monitor and enforce the extent to which SADA (and all water utilities) meets the three complaint targets

---

[98] Personal interview with SADA's Information Centre supervisor, 1 Aug. 2010.

[99] Personal interview on 9 March 2011.

[100] See Appendix 8.

it sets out in Table 5.13 (SADA's targets are shown in Table 5.13). One can foresee that the reform will drive government-owned water utilities SADA to pay more attention to customer complaints.

Table 5.13. SADA's complaint targets for 2011-2013.[1]

| Type of complaints | Targets baseline (2010) | 2011 | 2012 | 2013 |
|---|---|---|---|---|
| Billing complaints: % responded to within 3 working days | 100 | 100 | 100 | 100 |
| Response to complaints: % complaints meaningfully responded to within 5 working days | 92 | 95 | 97 | 98 |
| Telephone complaints: % responded to within 30 seconds | 95 | 97 | 98 | 98 |

[1] Based on data which SADA provides to the NWSC.

### 5.4.6 Conclusions of the case study

Overall, this analysis reveals that the operational efficiency of water utilities had no direct relationship with the reform. Two in-depth case studies (of PBAPP and SADA) further re-affirms that the reform has not influenced the operational efficiency of private (PBAPP) and public (SADA) water utilities. It gives a clear indication that the operational efficiency of water utilities depended on the conditions that existed before the reform was introduced. These case studies also show that the new conditions introduced by the reform have not yet bought any significant changes in the performance of water utilities. In terms of the research as a whole, these case studies have answered one of the central research questions (Question 3).

It will be a considerable amount of time before the results of the reform become visible. What really matters now is the continuity of the reform process. However, both case studies show that in certain areas – NRW and collection efficiency – water utilities have registered efficiency gains, even if there is no concrete evidence linking these to the reform. Notwithstanding all this, one can expect that the reform will drive water utilities to achieve higher efficiency not only in those areas where they have already excelled, but also in low efficiency areas – production cost and complaint management. In fact, mechanisms and approaches such as regulation, corporatization, benchmarking, and flexible financing have been laid down to encourage and structure efficiency gains among water utilities. Nevertheless, the effectiveness of these mechanisms and approaches are yet to be seen.

## 5.5 Conclusions

This chapter has analysed the effects of the reform on the performance of water utilities, using four indicators: non-revenue water, collection efficiency, unit production cost and customer complaints. These indicators are collectively seen as reflecting operational efficiency. Comparison was made between public and private water utilities, and was furthered refined through the two case studies. Three conclusions can be drawn from this chapter.

The first one is related to the effect or contribution of the reform in improving the utilization of the performance indicators for the Malaysia water sector. The chapter started out by suggesting four indicators which might be useful for measuring the operational efficiency of water utilities. Other indicators could later be included. Following the institutional reform, which led to the establishment of the central regulator – the NWSC – and the Water Forum, it is assured that it's systematically implementation can be enforced and monitored. These (standardised) indicators for all water utilities replace the old ones which were previously implemented by the individual water utilities.

Secondly, it can be concluded that the private water utilities have proved themselves to be more effective than their counterparts in the public sector in managing water losses and in collection efficiency. There are no clear differences in the performance of the two types of company with regard to cost management and managing customer complaints. As already explained, it is too early at this juncture to equate the changes in operational efficiency solely to the reform. The two in-depth case studies further re-affirm the finding that the operational efficiency of water utilities was primarily influenced by the conditions that existed prior to reform. The Malaysian water sector is a patchwork of private and public set-ups, the operational efficiency of which, to certain extent, reflects differences in terms of their operations, ownerships, etc. This strongly suggests that a 'one size fits all' solution is not tenable. For instance, private water utilities are more financially able to effectively address water losses than the public sector. As such, it is critical that the reform process recognizes these differences. In addition, the comparisons that have been made in the performance of public and private utilities indicate the importance of the availability of (reliable) information. This leads us to a third conclusion, concerning the role of (reliable) information in effective regulation.

The third conclusion is that the availability of (reliable) information is one of the most important prerequisites for effective regulation. In fact all the operational efficiency indicators are measured on the basis on information which then forms the basis for regulation. The analysis shows two main challenges about this information which the central government needs to address. First, acquiring the necessary (quality) information is a challenge for the central regulator. The case studies indicated that, while information is readily available from the private water utilities, it is not an easy task to get the same (quality of) information from public water utilities and this might be one of the most difficult tasks facing the central regulator. Without quality information, it is difficult for the NWSC to formulate accurate performance indicators for the water sector. The requirement for third party validation of information (under Section 29 of the WSIA) would do much to enhance the reliability and accuracy of information provided. Another question that needs addressing is how the information asymmetry, especially between the information-rich (private) water utilities and the central regulator, can be reduced.

# Chapter 6.
# The environmental effectiveness of the water supply sector

*'We do not inherit water from our grandfathers,*
*but we borrow it from our next generation.'*

Kenyan proverb

*'Only when the last tree has been cut down,*
*Only when the last fish has been caught,*
*Only when the last river has been poisoned,*
*Only then you realize that money cannot be eaten.'*

Cree Indian prophecy (taken from Chan, 2007)

## 6.1 Introduction

The above proverbs clearly express the need to value the world's precious water resources and safeguard them for the benefit of future generations. Asian countries face great difficulties in meeting this challenge. One third of the Asian population does not have access to a regular supply of safe water (Asian Development Bank, 2006) and around two thirds of the people in the world with limited or no access to water lives in Asia. The Asian Development Bank also estimates that countries in Asia (and the Pacific) would need at least US$ 8 billion to meet Target 10 of the UN's Millennium Development Goals by 2015. Although WHO and UNICEF more recently (2010) predicted that this target would probably be met by 2015, still there are 672 million people in the world who lack access to safe drinking water.

The water sector is closely related to the environment. It is the environment – rivers, lakes, canals, catchment areas – that provides raw water for water companies to treat and sell to consumers. It is also the environment which receives used water from consumers. With more and more rivers being polluted as a result of rapid urbanization and industrialization, there is increasing pressure on the availability of fresh water. Water companies in Malaysia have started to take the first steps towards improving their environmental performance, recognizing the need for them to promote harmonious co-existence between economic and environmental considerations. Governments should support the companies in these efforts and their water policies should aim to strike a balance between environmental and economic interests and between ecological and economic rationality (Mol, 1995).

It is expected that environmental considerations will gain wider recognition and increased momentum in Malaysia's economic planning, including its water supply sector (Economic Planning Unit, 2010). For instance, the current 10th Malaysia Plan (2011-2015) emphasizes developing environmental resources in a sustainable manner (Economic Planning Unit, 2010). In this chapter I explore how the water sector is responding to the government's policy on environmental issues. This is of key concern as in the past the water sector has been adopting environmentally unfriendly practices in areas such as sludge management (PAAB, 2009). This chapter not only examines sludge management but also other aspects of environmental performance: compliance with drinking water standards and information disclosure.

At the moment there are no specific indicators in place to measure the environmental effectiveness of the water sector in Malaysia. This chapter proposes that these three indicators are useful for measuring this. The main emphasis is on analyzing the water sector's performance with regard to these indicators, and the extent to which the reform has (or has not) affected this performance. Wherever applicable, the analysis is made by comparing the period before and after the water sector was reformed. However, full comparisons are not always possible due to data scarcity. Section 6.2 briefly describes the methodological approach to the research. Section 6.3 assesses the general performance of the water sector against these three environmental indicators. Section 6.4 presents an in-depth case study of two companies, comparing the performance of a private water utility (PBAPP) with that of a public water corporation (SADA). Section 6.5 sets out the conclusions.

## 6.2 Methodology

### 6.2.1 Data sources

Data for this chapter were generated through three main sources: in-depth interviews with key informants, surveys and secondary data collection.

In-depth interviews were conducted with four groups of stakeholders: (1) water utilities (public[101] and private); (2) officials from public authorities – the Ministry of Energy, Water and Communications[102] (MEWC), the Economic Planning Unit of the Prime Minister's Department (policy maker), the National Water Services Commission (NWSC) (industry regulator) and the Department of Environment (DOE) (responsible for enforcing the EQA 1974); (3) consumer associations; and (4) environmental organizations. In total, 53 interviews were held from April 2009 to August 2010 (see Appendix 9 for list of interviews). This included 13 interviews conducted during the course of the two in-depth case studies with PBAPP and SADA. Interviews were used to get answers to 'why' rather than to 'what' questions (Yin, 2009). They were also used to ascertain qualitative data for all indicators. A summary of the interviews is presented in Table 6.1. (Direct) observations and visits to water treatment plants, reservoirs and water labs were also carried out during the two in-depth case studies. It gave the researcher the opportunity to observe the availability of the sludge treatment facilities and how water utilities handled sludge management – handling and disposal – and to observe the state of water labs and what they were capable of.

Surveys were used to obtain quantitative data, mainly from water utilities. These data were meant to complement the qualitative data gathered through the interviews. The quantitative data collected through the surveys were related to 'what' questions on the three indicators analyzed in this chapter. A total of 35 questionnaires with open-ended and closed questions were distributed to 20 water utilities from April 2009 to August 2010. Three different questionnaires were used: the

---

[101] Include state corporatized water utilities.

[102] From March 2008, it is known as the Ministry of Energy, Green Technology and Water.

*Table 6.1. Summary of interviews.*

| Stakeholders | Number of interviews |
| --- | --- |
| Water utilities | 38 |
| Public authorities | |
| • MEWC | 1 |
| • EPU | 1 |
| • NWSC | 3 |
| • DOE | 2 |
| Consumer associations | 4 |
| Environmental organizations | 4 |
| Total | 53 |

first for treatment only operators (8 questionnaires); the second set for distribution only operators (2 questionnaires); and the third set for treatment and distribution operators (25 questionnaires) (see Appendix 10). The surveys were administered after the in-depth interviews. Respondents were asked to return the questionnaires, using a pre-stamped envelope. Twenty one (60%) completed questionnaires were returned (Table 6.2). For the purpose of the analysis, the term 'public' will be used to refer to both state water departments and corporatized water utilities.

Other secondary data were sourced from water utilities' annual reports[103] and publications, legal statutes (such as the WSIA 2006, SPANA 2006 and EQA 1974), official government publications (such as the 9th and 10th Malaysia Plans and the National Water Resources Study), and the Malaysia Water Industry Guide published by the Malaysian Water Association.

*Table 6.2. Summary of surveys.*

| Water utilities | No. of questionnaires issued | No. of questionnaires returned | Percentage of questionnaires returned |
| --- | --- | --- | --- |
| Public | 8 | 4 | 50% |
| Corporatized | 10 | 2 | 20% |
| Private | 17 | 15 | 88% |
| Total | 35 | 21 | 60% |

[103] Only produced by water utilities listed on the Malaysian Stock Exchange.

## 6.2.2 Data analysis

This section explains how the indicators of environmental effectiveness – sludge management, compliance with drinking water standards and information disclosure – are analyzed.

Four specific aspects of sludge management were identified for analysis. The first analysis examines whether water utilities have on-site sludge treatment facilities, as required under the 2005 Environmental Quality (Scheduled Waste) Regulation (Department of Environment, 2005). This regulation categorizes sludge from water treatment as a 'scheduled waste' which must be treated on site or disposed at prescribed premises. At the moment, all scheduled waste substances are required to be treated and disposed of either at Kualiti Alam Sdn Bhd,[104] a private scheduled waste facility in Bukit Nenas, Negeri Sembilan, or at the DOE's approved sanitary landfill or disposal sites. The second analysis investigates the adoption of environmentally-friendly sludge treatment facilities by water utilities. The third analysis looks at sludge recycling and re-utilization initiatives. The fourth and final analysis examines water utilities' implementation of environmental activities, such as an environmental policy, charter or pledge, and their reactions to the idea of a green tax. The analysis of the last three aspects is not based on any regulations, since there are none in place. However, they are all critical indicators for measuring environmental performance, and there are proposals for incorporating them in future reforms.

The National Guideline for Drinking Water Standard 2001 (NGDWS) issued by the Ministry of Health will be used as the premise for the second indicator, which measures the performance of water utilities in relation to drinking water quality (Ministry of Health, 2008). The NGDWS identifies five physical parameters for measuring water quality standards: (1) residual chlorine; (2) faecal coliform; (3) *E. coli*; (4) turbidity; and (5) pH (Table 6.3).

Another aspect of water quality is the readiness of water utility companies to comply with Section 41 of WSIA. Section 41 requires 'water distribution licensees' to comply with minimum quality standards (as prescribed by the Minister) when supplying water to any premises (MEWC, 2006a: 40).

The final indicator – information disclosure – is a new indicator, introduced in this research. It has never been used before to measure environmental performance amongst water utility

*Table 6.3. Mandatory water quality standards (Ministry of Health).*

| Compliance parameters | Mandatory standards |
|---|---|
| Residual chlorine | 0.1 ml/l |
| Faecal coliform | absent |
| E. coli | absent in 100 ml sample |
| Turbidity | 5 nephelometric turbidity units |
| pH | 6.5-8.5 |

---

[104] Further information about the company can be found at www.kualitialam.com.

companies in Malaysia. However it is anticipated that it will be included as an essential ingredient of future reforms (SPAN, 2010b). Two areas of information disclosure are analyzed. First I look at the extent to which water utility companies record and report their compliance with sludge management and water quality standards. Second I look at how water utility companies have prepared themselves to comply with Section 29 of the WSIA, which obliges them to provide information to the central regulator, NWSC (MEWC, 2006a).

The summary of the indicators used for analysing environmental effectiveness is shown in Table 6.4.

*Table 6.4. Summary of indicators used to analyse environmental effectiveness.*

| Indicators | Areas of analysis |
| --- | --- |
| Sludge management | Presence of on-site treatment facilities |
| | Adoption of environment-friendly treatment facilities |
| | Sludge recycling and re-utilization |
| | Environmental concerns |
| Compliance with drinking water standards | Compliance with NGDWS 2001 |
| | Readiness to comply with Section 41 of WSIA |
| Information disclosure | Recording and reporting on sludge management and water quality compliance |
| | Readiness to comply with Section 29 of WSIA |

## 6.3 Water reform and environmental effectiveness

### 6.3.1 Sludge management

Sludge is a by-product of the water treatment process. It consists of the substances removed from raw water, and the agents added to raw water during coagulation and filtration (Makris & O'Connor, 2007). Direct discharge of sludge into water courses, such as rivers, lakes and canals, pollutes these sources, reducing the availability of water resources for future utilization. The presence of toxic materials such as aluminum and arsenic in sludge poses a serious threat to raw water quality (Makris & O'Connor, 2007). The issue of sustainable sludge management in Malaysia is becoming more pressing in view of the increasing amounts of sludge being produced by water treatment plants throughout the country. For instance, the Langat 2 water treatment plant is estimated to generate around 400-500 tonnes of sludge per day in Selangor (PAAB, 2009), while another 600 tonnes of sludge is generated daily by the 29 water treatment plants managed by PNSB.[105] The increase in the amount of sludge is largely caused by the increase of water treated to cater for

---

[105] PNSB is the treatment operator in the State of Selangor.

growing water demand in Malaysia. This has risen from 9,666 million litres per day (MLD) in 2000 to 15,285 MLD in 2010. Water demand is expected to reach 20,338 MLD in 2020 and 31,628 MLD in 2050 (Economic Planning Unit, 2000).

The Environment Quality Act (EQA) 1974 is Malaysia's main legal framework for regulating sludge management and other hazardous substances. The EQA 1974 classifies sludge as a 'scheduled waste' due to the presence of heavy metals in it. As such, it has to be properly treated at designated sites prior to disposal.

This section analyzes three aspects of sludge management. The first is the availability of on-site sludge treatment facilities and the adoption of environmentally-friendly sludge treatment facilities. The second aspects concerns initiatives undertaken by water utility companies to recycle or reuse the sludge, and incentives (from government) to facilitate such initiatives. Finally, I examine the extent to which water utility companies take environmental considerations into account in their daily operations.

## The availability of on-site sludge treatment facilities

The survey[106] conducted among water utility companies (both public and private) revealed that 72% of the water treatment plants lack on-site sludge treatment facilities. The majority of these plants – almost 80% – were owned by public water departments (Table 6.5). These plants directly discharge sludge into rivers even though this clearly contravenes Regulation 4 of the Environmental Quality (Scheduled Waste) Regulation 2005. This result is in line with findings from the drinking water quality study conducted by the National Audit Department (Jabatan Audit Negara, 2008).

Sludge lagoons were the preferred treatment method among the 27 plants that were equipped with on-site treatment facilities. A total of 24 plants (89%) used this method. Other treatment methods include sludge recovery tanks for making sludge cake, drying beds, a dewatering centrifuge or flat-sheet membranes. These technologies only prevent the direct discharge of sludge into waterways. Heavy metals are still present in the settled sludge, and a representative from the DOE stated that 'some form of control is needed to ensure sludge is properly treated prior to disposal'.[107]

*Table 6.5. The availability of on-site sludge treatment facilities.*

| Ownership | WTP managed | WTP with on-site treatment facilities (n) | % | WTP without on-site treatment facilities (n) | % |
|---|---|---|---|---|---|
| Private | 63 | 13 | 21 | 50 | 79 |
| Public | 33 | 14 | 42 | 19 | 58 |
| Total | 96 | 27 | 28 | 69 | 72 |

---

[106] Based on 60% (21/35) of returned questionnaires.

[107] Personal interview on 16 Oct. 2009.

Questions were also asked about the constraints that prevent water utility companies from installing treatment facilities, and where they dispose of the sludge. It was obvious that the lack of treatment facilities, meant that almost all water utility companies discharged (raw) sludge back into the environment from which drinking water is obtained (Jabatan Audit Negara, 2008). Direct discharge of sludge may be the cheapest solution for water utilities, but comes at a high price for consumers and the environment. Heavily polluted rivers not only threaten aquatic life, but are also likely to contribute to increased water treatment costs as additional doses of chemicals are needed at the coagulation and flocculation stages. This in turn will lead utility companies to demand higher tariffs from consumers and increase the number of polluted rivers. While the DOE has an initiative to get people to regard rivers as something valuable which 'must be treated as common goods rather than common waste'[108] a representative of the NWSC believes that the government 'is politically not ready to include water resources in the (water supply) reforms as water resources are a matter for state governments'.[109] This respondent thought that the federal government did not want to encroach into state government matters as this would jeopardize cordial federal-state relationships. It is unlikely that jurisdiction over water resources will change under the current political set-up.

Table 6.5 shows that the majority of water treatment plants (public and private alike) lack sludge treatment facilities. It can be concluded that the reforms have not as yet had any significant influence on this matter. One of the reasons given by water utility companies for not having on-site sludge treatment facilities is that most, if not all, of these plants are old and were built long before the water sector was reformed. The companies acquired them from the Public Works Department or state governments in the 1950s and 1960s (through various modes of acquisition/ asset transfer/private participation). At that time the public water utilities were not obliged to have sludge treatment facilities. Moreover, building sludge treatment facilities and disposal sites would require sizeable areas of land, and most of the water utility companies do not have sufficiently large land banks. Finally, the utilities pointed out that the contracts that they signed with the state governments did not stipulate any obligation to install sludge treatment facilities. The utilities argued that it was the state government's obligation to build or provide treatment facilities if they were needed. Utilities did not have the finance to acquire the land needed to build sludge treatment facilities and if they had to raise this then they would have to pass on the extra cost to their customers.

Water utility companies were also concerned about the high transportation costs if they were obliged to dispose of large quantities sludge at recognized disposal sites. Sometimes this could involve trips of up to 100km. Under the present situation, which a representative of a private water company described as one where 'there are no such disposal sites available close to the locations where we operate',[110] it is not surprising that many companies feel that their only option is to discharge sludge directly into water courses. However, the water utilities agreed that a permanent solution to this problem must be sought. The General Manager of the Taliworks Corporation – a private water operator in Kedah – suggested that 'sludge issues could be managed in a more

---

[108] Personal interview on 2 Oct. 2009.

[109] Personal interview on 23 Sept. 2009.

[110] Personal interview on 8 May 2009.

sustainable way under one single authority'.[111] He argued that disposing sludge at landfill only provides a temporary solution as the day will come when the existing landfill can no longer take in the growing amount of sludge for disposal. He thought that sludge re-cycling and re-utilization were the only sustainable solutions for sludge management. These approaches not only minimize the amount of treated sludge, but also convert it into useful products and can therefore also generate income for the water utilities.

The lack of available sludge treatment facilities was a problem that existed long before the idea for the reform was mooted. It remains a problem for both public and private water utilities. Financial considerations and a lack of availability of space or land are the main factors preventing water utilities from deploying treatment facilities.

## *The adoption of environmental-friendly sludge treatment facilities*

Generally, the majority of the water utilities indicated their readiness to adopt environmentally-friendly sludge treatment facilities for their water treatment plants in the future. These technologies include decantering, membrane filtration, drying beds and others. However, they did not reveal any specific time frame when they might be ready to adopt these technologies.

Cost seems to be the main obstacle which preventing water utilities from adopting greener sludge treatment technologies. Nevertheless Salcon Engineering – a private operating and maintenance water operator in Negeri Sembilan – indicated that it would consider deploying 'green technology' in their plants if its contract was renewed. The company was looking at adopting 'a cyclone system or decanter to dump the sludge because of limited available space. But as private utilities, we are concerned about dollars and cents'.[112]

Gamuda Water – a private water operator in Selangor – was concerned about the high water tariff that might result if they were to acquire environmentally-friendly sludge treatment facilities. Its General Manager said that if water utilities had to bear these costs, this would force them to ask for a higher bulk supply rate, which would eventually be passed on to the public in terms of higher water tariffs'.[113] An alternative, that avoids any unnecessary tariff increase, would be for the government to provide financial assistance to help water utilities to meet the sludge disposal regulation, especially as some technologies – such as a mechanical sludge treatment system – can be very costly.

According to Syabas – the private water operator in the State of Selangor and the Federal Territory of Kuala Lumpur and Putrajaya – deploying environmentally-friendly treatment facilities is the only way to avoid the direct discharge of raw sludge into the streams. But this would not reduce the amount of sludge produced. So, according to the Executive Director, by solving one problem (adopting green sludge technology) 'we are actually creating another problem, which is where to dispose of the treated sludge'.[114] These concerns are valid. There is limited available land for dumping settled sludge and this problem will become more pressing given the anticipated 3.3%

---

[111] Personal interview on 8 May 2009.

[112] Personal interview on 25 May 2009.

[113] Personal interview on 27 May 2009.

[114] Personal interview on 23 July 2009.

increase in water demand by 2020 (Economic Planning Unit, 2000). Thus, it makes far more sense for any regional sludge treatment facilities to convert sludge into useable products, thereby reducing the amount of sludge that needs to be dumped. This would not only extend the usable life span of the existing landfill sites but allow some to be decommissioned and/or used for other purposes.

To conclude, water utilities do show an interest in deploying 'green' sludge treatment technologies, but their enthusiasm is dampened by the cost factor. In the long run it will not be an option for the utilities to externalize their environmental costs in this way. The new financial mechanisms being established by PAAB could make a significant contribution to accelerating the adoption of environmentally-friendly sludge management in the water sector.

## Sludge recycling and re-utilization: converting waste to wealth

In Malaysia, sludge recycling or re-utilization is a new area within the water sector and its potential benefits are yet to be explored. However, several studies conducted at the university level have indicated the potential uses of sludge. For instance, Wahid et al. (2008) revealed that sludge has plasticity characteristics that allow it to be shaped and molded into pottery products. Hassan (2006), Wan Jusoh (2007) and Syed Zin (2007) have also studied the potential use of sludge in ceramics. In most developed countries, sludge recycling has been extensively promoted as an environmentally-friendly disposal method. Makris and O'Connor (2007) argue that sludge recycling is not only environmentally-friendly, but also has cost reduction advantages, since less sludge contamination of streams results in lower costs for drinking water treatment.

Many studies have also pointed out other potential uses for sludge. These include land application (Ippolito, Barbarick & Elliot, 2011; Brinton, O'Connor & Oladeji, 2008; Agvin-Birikorang, Oladeji, O'Connor, Obreza & Capece, 2009; Novak & Watts, 2005; Walsh, Lake & Gagnon, 2008; AWWARF, 2007), brick manufacturing (Iacob & Farcas, 2010; Huang, Pan, Sun & Liaw, 2001; Tay, Show & Hong, 2001; Hsieh & Raghu, 2008), land reclamation (Basta & Dayton, 2001; Hsieh & Raghu, 2008; AWWARF, 2007) and cement production (Hsieh & Raghu, 2008). In the Netherlands, in 2009 99.8% of the sludge generated from treating drinking water is recycled (VEWIN, 2010). In 2006, this figure was 94% (VEWIN, 2006). Water companies in the Netherlands have jointly established the Residues Union to spearhead sludge recycling and to explore potential uses of sludge. At present the recycled sludge produced by Dutch water companies is widely used for brick making, materials for road barriers and foundations, land elevation and ballast material for industrial use (VEWIN, 2010).

At present the reform of the water sector has not had any great influence on sludge recycling. The survey revealed that the majority of water utilities did not think it was feasible at present to convert sludge into usable products, although some had explored this option. In general, the private water utilities showed more initiative in this respect than their counterparts in the public sector. Syabas, ABASS[115] and PBAPP had all attempted to convert sludge into bricks, pottery products and pellets for power generation, but their efforts were hindered by two factors, which make sludge recycling unpopular. First, they did not see the 'business sense' for taking such initiative, which was both costly and showed no economic return. A representative from ABASS revealed

---

[115] ABASS is the treatment operator in Selangor.

that, 'it is economically not viable to convert sludge into bricks as it shrinks by about 40%, and has no bonding properties. The process also involves additional costs in transporting the sludge and acquiring clay to mix with it'.[116] Moreover, water utilities are not legally required to recycle sludge and thus see no need to spend money on something unprofitable that had no demand.

In addition to problems of feasibility and cost water utilities were of the opinion that the classification of sludge as a 'scheduled waste' under EQA 1974 implied that sludge converted into other products was not safe for application, and reduced the public's acceptance of products made from recycled sludge. This led the water utilities to urge the DOE to consider declassifying sludge as a 'scheduled waste'. They cited several studies which revealed that sludge from water treatment did not exhibit the characteristics of a scheduled waste and did not warrant being classified as such (Aminudin, 2009; PAAB, 2009). The Chief Operating Officer of ABASS – a private operator in Selangor – pointed out 'the ammonia released by the Indah Water Konsortium's sewerage plants is not classified as 'scheduled waste' even though it clearly contravenes the law'.[117] The DOE, however, do not see the need to amend the law to accommodate the request from water utilities. The Deputy Director-General of the DOE is convinced that 'some form of control is needed due to the presence of heavy metals in the sludge. We believe that sludge needs proper treatment before disposal. However, water utilities can apply for exemption from this regulation under the Guideline for Application of Special Management of Scheduled Waste, which allows them to dispose sludge at sanitary landfill sites'.[118]

Given the reluctance of Malaysian water utilities to undertake sludge recycling, the following paragraphs look at the role the government could play in facilitating sludge recycling. At present, sludge recycling is not mandatory, and is occasionally carried out on a voluntary basis by water utilities. However there is no authority that has taken up the role of promoting and coordinating sludge recycling among water utilities (and other industries). The water utilities think government could play a leading role in promoting sludge recycling.

The interviews revealed a general concern among water utilities about the possible environmental threat and recognition of the need for proper sludge treatment and disposal. Water companies believe that the reform could be used to pave the way towards sustainable sludge management and saw three possible options: setting up regional sludge treatment companies; providing financial incentives for acquiring environmentally-friendly sludge technologies; and promoting research and development in sludge recycling.

Water utilities believed that the government should consider setting up a regional sludge treatment company (RTC) to facilitate and coordinate sludge recycling among water utilities. They cited the Residue Union in the Netherlands as an example. They thought a similar body could be set up in Malaysia as a government-linked company which should be managed as a business entity. Another possible option suggested by a representative from PBAPP is 'for the government and water utilities to jointly sponsor the establishment of the RTC'.[119] The water utilities thought that a RTC should initially focus on research and development into the beneficial uses of sludge

---

[116] Personal interview on 19 June 2009.

[117] Personal interview on 19 June 2009.

[118] Personal interview on 16 Oct. 2009.

[119] Personal interview on 4 Nov. 2009.

and later on production and commercialization. Such a proposal fits nicely into the broad overall policy direction towards holistic water management, providing a link between the water sector and the sewerage sector (MEWC, 2008). These two sectors should not be segregated, as both are inextricably linked in the water cycle chain. The CEO of NWSC saw the link between the two and proposed that the RTC should also consider 'extending their scope of work to include sewerage (and waste water) sludge in the future'.[120]

While establishing a RTC is quite a long-term solution, water utilities are interested in installing environmentally-friendly sludge treatment technologies in their plants. However, the cost factor was preventing them from acquiring such technologies or equipment. Their concern is substantiated by the study conducted by the national water asset management company, PAAB (2009), which showed that high investments in acquiring such technologies would result in higher treatment costs and higher water tariffs. Because of this the water utilities want the government to provide financial incentives, such as subsidies or grants, to promote the adoption of 'green technologies' in sludge management. It is possible that the government can extend existing economic instruments – as has been recommended in Economic Planning Unit's Handbook for Economic Instruments for the Environmental Management for Sludge Management – to sludge recycling (Economic Planning Unit, 2004). For instance, a deposit-refund system and revenue neutrality could be used to encourage investment in environmentally-friendly sludge technologies by water utilities. Water users are also expected to favour this option as this will not burden them with extra costs. Another form of incentive that the government might consider is environmental taxes (in the form of pollution or product taxes), the revenues from which could be used to mitigate the effect of the direct discharges of sludge into the environment (Economic Planning Unit, 2004). Such taxes would also act as disincentive to directly discharge sludge into water courses. However, the government will have to be mindful of the financial implications of such taxes, as these will result in higher operating costs for water utilities. In most cases these extra costs will be passed on to water users through higher water charges. Moreover, imposing such taxes might also have significant socio-political implications. Socially, they will generate a lot of resistance from water users, and local politicians may see them as diluting political support among their constituents.

The PBAPP has explicitly called on the government to facilitate research and development into sludge recycling. Studies by the MWA (2008c) and PAAB (2009) have emphasized the importance of having coordinated research and development activities, especially to identify new potential uses for sludge. The PBAPP has suggested that the government impose a mandatory requirement for every water utility to allocate certain percentage (they suggested 2%) of its revenues for research and development activities. Another interesting proposal from the Strategic Planning Manager of PBAPP was for the NWSC 'to internalize the costs for sludge treatment as part of the environmental costs, and allow this cost to be reflected in the tariff revision'.[121]

---

[120] Personal interview on 23 Sept. 2009.

[121] Personal interview on 12 Nov. 2009.

*Environmental concerns*

This section analyses two environmental concerns among water utilities. First, it examines water utilities adopting environmental pledge to act as guiding principles in their day to day operations. Here we look at two types of environmental pledge; those covering policy and practices. Second, the analysis examines the utilities' attitudes towards green taxes as an economic instrument that could conserve water resources and encourage water conservation.

*Environmental pledges*

The research clearly demonstrated that private water utilities show more concern for the environment when conducting their water business activities than public ones. They almost universally document and publish their environmental policy or practices. The analysis here is based on information gathered from water utilities' annual reports. It is worth noting that these are only published by the private sector, while none of the public water departments published annual reports.

Data gathered from annual reports of five private water utilities[122] showed that they all contain a special section that reports on their environmental commitment. This can be seen as a move towards complying with the requirement for information specified under Section 29 of the WSIA (MEWC, 2006a). We will investigate the extent to which these water utilities consider and report on their environmental commitment.

The outstanding example is PBAPP, which in many respects is one of Malaysia's leading water utilities. PBAPP's commitment to environmental issues is clearly stated in its environmental policy, which reads as follows:

> 'PBAPP is fully committed towards protecting, preserving and conserving the environment while striving to meet all of Penang's water needs.'

PBAPP's commitment towards prudent environmental management is internationally recognized. Three of its facilities – the Batu Ferringi and the Waterfall Treatment Plant, and Teluk Bahang Dam – received international ISO 14001 certification for their environmentally-friendly water supply management system (in 2005 and 2007). The company is seeking to secure similar certification for two more of its facilities: the Air Itam Dam and the Rifle Range One Stop Operations Centre.

PNSB's Annual Report (2008) sets out its vision of corporate responsibility: '(Puncak Niaga Sdn. Bhd)[123] is mindful of environmental preservation as we journey on as a society and a nation to achieve the agenda of attaining the aspirations of Vision 2020'. In practical terms, PNSB established a Water Resources and Environmental Surveillance Department (WRES) in 1995, to provide effective environmental services and consistent delivery of high quality drinking water. WRES undertakes special environmental investigations of water catchment areas, investigations of violations and reports raw water pollution to the relevant authorities such as DOE and LUAS

---

[122] The others do not publish annual reports.
[123] PNSB is treatment operator in Selangor. It holds 70% equity in Syabas.

– the water resources regulator – so that they can take the necessary action. PNSB is also active on the educational and awareness raising fronts and has formed the River Rescue Brigade, which aims to educate the younger generation on the importance of environmental preservation and conservation.

Another private utility, SAJH[124] recognizes that good environmental practice is the key to 'ensure the viability of operations for the long term' (Ranhill Utilities Berhad, 2007). Its good environmental practices include mitigating the effect of climate change (floods and drought) on its operations, using TNB's[125] environmentally-friendly power supply (as opposed to diesel power) and ensuring proper treatment of sludge prior to disposal. SAJH sees their efforts as making a contribution to the UN Millennium Development Goal (Goal 7) that calls for nations to promote environmental sustainability.

Salcon Engineering's environmental policy is 'to minimize negative environmental impacts as well as to promote environmental conservation in its business operations' (Salcon Engineering, 2008). A representative of Salcon said that 'the company is a responsible operating and maintenance operator that is committed to conducting environmental impact assessments for its water development projects to ensure that the projects harmoniously co-exist with the surrounding ecosystem.'[126] It is also high on Salcon's agenda to ensure that sludge generated from its water treatment plants is disposed of in an environmentally responsible manner at Kualiti Alam, a dedicated disposal site for scheduled waste substance.

Aliran Ehsan Resources Berhad also recognizes 'the importance of a clean and safe environment and occupational health and safety practices' (Aliran Ehsan Resources Berhad, 2008). In this respect the company says it annually cleans the rivers and drains within the vicinity of its treatment plants to avoid its operations causing any unnecessary environmental disturbances.

The above examples show how private water utilities report their environmental concerns. At the moment only private water utilities have established and published their commitment to the environment. The absence of such practices amongst public water utilities is a cause of concern. It indicates a lack of awareness of the potential threat that the water sector poses to the environment, and of the balance that the water sector needs to strike between environmental and economic considerations.

## Attitudes towards green taxes

Many countries have used market-based mechanisms to influence the behaviour of utilities, such as the water sector (Economic Planning Unit, 2004). For example, incentives (e.g. subsidies) are commonly used to reward desired behaviour; while taxes are used to penalize undesired behaviour. Thus a 'carrot and stick' approach is applied. This is often based on the principle that polluters must bear the cost for cleaning up the pollution they have created. This can also be extended to encouraging water conservation. For instance, in Singapore households face water conservation

---

[124] SAJH is distribution operator in Johor.

[125] TNB is a power generating and supply company.

[126] Personal interview on 25 May 2009.

charges if they consume more than 40 cubic metres of water per month (Khoo, 2007). Similar water conservation surcharge is currently being enforced in Penang (PBAPP, 2009).

Subjecting water users to paying the costs of treatment of the (waste) water they used can be done by incorporating a 'green tax' into the water tariff structure. Generally, a green tax (or environmental tax) can be broadly defined as a (monetary) amount charged to consumers' bill for the consumption of certain services which can degrade the environment. In the water sector, these taxes can be used to treat waste water and sewerage (as a result of water consumption) and for the purpose of river conservation. This is based on the principle that the more water used, the more waste water is produced, thus more money needed to treat the waste water. The interviews showed that the utilities see a 'green tax' as a good and workable solution, but consider that its implementation might be a problematic at this point in time.

Most water utilities are convinced that the incorporation of a 'green tax' as (a small) part of the water tariff would be a good move towards environmental protection. They agreed that the revenues generated from this tax could be channeled to assist state governments to conserve rivers, vital for ensuring an adequate supply of raw water for future use. However, they did not want state governments to use this mechanism as a means of getting additional revenues. The General Manager of the MUC – a private water treatment operator in Perak – felt that the state government should be more motivated 'to protect catchment areas as it receives royalties from selling raw water, and should employ all means to prevent any uncontrolled economic activities in the catchment areas'.[127] A representative from GSL Water stressed the need for state governments to be transparent and accountable in administering these revenues and, most importantly, 'the public must be able to see the benefits of paying tax'.[128] Utility companies thought there was a clear role for the NWSC to regulate how these revenues are used.

While accepting a 'green tax' was a good move, water utilities felt that its implementation could be problematic, for at least two reasons. First, imposing a 'green tax' would mean consumers paying more for water, which might not be popular. Chan (2007) argues that even though water bills represent about 10% of what people pay for electricity, any increase could decrease people's willingness to pay (for water), especially if consumers feel that they are not benefitting from what they are paying for. Secondly, the water utilities questioned whether imposing a 'green tax' could be justified while the state governments are so ineffective in enforcing existing environmental regulations over activities such as sand dredging. A representative of PNSB – a private water operator in Selangor – said the company was 'tired of the inaction from a government authority in tackling the issue of sand dredging'.[129] The Director of JBA Kedah anticipated that it might take '20-30 years before Malaysians are willing to accept environmental taxes as part of their water bill'.[130] As an alternative, a representative of Syabas – a private water distribution operator in Selangor – urged state governments (in collaboration with the NWSC) to take a bold step in setting proper (economic) tariffs as, in his words, 'the time has come for consumers to value water,

---

[127] Personal interview on 4 May 2009.

[128] Personal interview on 13 May 2009.

[129] Personal interview on 18 June 2009.

[130] Personal interview on 11 May 2009.

and realize that it's costly to bring water to their house'.[131] In addition, they wanted NWSC to promote the polluters pay principle as part of the bigger agenda in mitigating the effect of global warming on the water sector.

From the government's perspective there are many other issues that would need to be resolved before considering a 'green tax' on water. Even though one objective of the reform is to promote the adoption of single billing – where water and sewerage bills are combined – it will not be impossible to implement this in the near future. If implemented now, a representative of the Economic Planning Unit (of the Prime Minister's Department) feared that 'consumers might see this as an attempt (by the government) to boost the sewerage revenue (of Indah Water Konsortium)'.[132] It is also important to maintain awareness raising and educational programmes to bring about a change in public attitudes towards a 'green tax'. In the words of a representative from the MEWC:

> 'We have to change the perception of the public so they become aware that they are not just paying for water they consume, but also paying to protect and manage the environment. So, they are duty-bound to ensure that the environment is adequately protected.'[133]

He thought that several factors would need to be considered before a 'green tax' could be implemented in the water sector.

> 'I am of the opinion that a 'green tax' is good idea. The question is when. I see that political will is a prerequisite for its smooth implementation. We must also take into account acceptance from the public.'[134]

### 6.3.2 Compliance with drinking water quality standards

Section 41 of the WSIA requires water utilities to supply water to consumers which 'complies with the minimum quality standards prescribed by the Minister'[135] (MEWC, 2006a: 40). At the moment, the NWSC relies on the NGDWS 2001 issued by the Ministry of Health, although there are plans to introduce a new set of minimum quality standards. The NGDWS stipulates the limits for physical, chemical, microbiological and radiological parameters and all water utilities must comply with these standards (Ministry of Health, 2008). The NGDWS is enforced by the National Drinking Water Quality Surveillance Programme, which monitors the quality and safety of treated water and monitors and controls raw water sources and supplies. It provides an early warning signal to water utilities and health authorities if there is a need to take corrective actions to address problems with drinking water quality and safety, and health problems. Generally the

---

[131] Personal interview on 23 July 2009.

[132] Personal interview on 14 Oct. 2009.

[133] Personal interview on 17 Sept. 2009.

[134] Personal interview on 17 Sept. 2009.

[135] The Minister responsible for water supply.

monitoring is done at fixed intervals from specific sampling points throughout the supply system, from intakes at the treatment plant, through to the distribution system.

Section 41 of the WSIA is intended to make water quality regulation more stringent and coordinated. Compliance with minimum quality standards will be one of the key performance indicators imposed upon water utilities by the NWSC. However, it is expected that water utilities will be given sufficient time to take remedial actions before Section 41 is fully invoked. The ultimate goal is to ensure that all Malaysians have access to good quality water. This is not an over-ambitious goal and a representative from a consumer association is confident that when all the efforts are in place: 'one day we will be able to drink water from the tap'.[136] This section analyzes water utilities' compliance to the NGDWS and the steps taken to improve their compliance level.

*Compliance with the NGDWS 2001 and steps to improve compliance*

Generally, most water utilities have been achieving satisfactory levels of drinking water quality (Jabatan Audit Negara, 2008). Nevertheless, instances of non-compliance with the drinking water standards do occur. For instance, AWER (2011) reported that water in Kelantan was coloured and smelly. This section presents the results of a survey conducted amongst water utilities to determine their compliance levels with drinking water quality standards. It shows how compliance with (or violations of) drinking water standards is recorded. Compliance levels are divided into three broad categories: full compliance (or no violations detected), violations detected, and no compliance data available from (or not given by) water utilities.

Water utilities record their compliance levels in different ways. Some record them in absolute figures – quoting the numbers of violations per year – while others record them in percentage terms – measuring the numbers of failed samples in relation to the numbers of samples taken in one year. These differences in reporting are bound to lead to interpretation problems, not only by the regulator, but also by members of the public. In addition, this does not fit with the overall objective of promoting 'transparency in reporting of drinking water quality, and in the investigation of incidents affecting drinking water' (Rouse, 2007: 193). This led the NWSC to devote substantial time to developing a reliable and quality information gathering mechanism when it was first established. This goes to re-affirm the importance of information in ensuring effective regulation of the water sector.

Table 6.6 summarizes the compliance levels amongst 14[137] (3 public and 11 private) water utilities which responded to the surveys. Details of the compliance data can be found in Appendix 11.

It is interesting to note that no data were available from publicly owned companies. This is probably due to them having poor data management systems, which prevent them from providing such information. This means that we can only look at the performance of private sector water utilities. Most of the companies who replied to the survey were fully compliant with minimum water quality standards, which might very well be caused by the fact that fully complying companies find it easier to answer the survey. Such compliance is being imposed as one of the key performance indicators on private water utilities in their contracts with state governments.

---

[136] Personal interview on 3 Aug. 2009.

[137] Out of 21 water utilities responded to the survey, only 14 answered this section.

*Table 6.6. Levels of compliance with NGDWS 2001 for 14 water utilities (2005-2008).*

| Ownership | Fully compliant | Violation(s) detected | No data |
|-----------|-----------------|------------------------|---------|
| Public    | 0               | 1                      | 2       |
| Private   | 5               | 4                      | 2       |
| Total     | 5               | 5                      | 4       |

Source: surveys.

Failing to comply will cause the water utilities to be financially penalized. On the other hand, meeting this standard (or moving towards it) can increase their reputation in the eyes of state governments and regulators. This will enhance their relationship with state governments, opening up further business opportunities with the state in the water sector. These pressures have led private water utilities to invest heavily in infrastructure, as well as in water testing procedures that meet international standards. Taliworks, PBAPP and ABASS had their water labs accredited to ISO/IEC17025 standards by the Malaysian Department of Standards. All these labs are administered by at least one qualified chemist. The companies also regularly flush the distribution mains, clean the reservoirs and clear water tanks in order to ensure the quality of water supplied to consumers.

Four (out of 11) private water utilities responding to the survey – Gamuda Water, GSL Water and AIUB and Syabas – reported violations of the drinking water quality standards. Three of them recorded these violations in absolute numbers, while the fourth (Syabas) recorded them as a percentage of total samples.

Syabas has shown a steady improvement in reducing its non-compliance with NGDWS and has managed to reduce the percentage of the samples failing the water quality sampling test. Since taking over the water supply (from PUAS) in 2005, it has done reasonably well in maintaining water quality in Selangor and the Federal Territory of Kuala Lumpur and Putrajaya. This was part of the contractual obligation under the 30-year concession with the state government of Selangor, which obliges Syabas to invest RM 7.1 billion in asset replacement programmes, including pipe replacements. Syabas' customers can look forward to an improved water quality in the near future once the on-going consolidation and migration exercise is completed.

GSL Water noted violations that happened at their plants due to internal factors, although some violations were also due to very turbid raw water from the river, which was beyond their control. This affected the company's capability to produce treated water that meet NGDWS 2001 requirement. The company argued that since jurisdiction over rivers is with the state government, 'not much can be done (on the company's side) to rectify the problems'.[138]

Four (out of 14) water utilities did not furnish any drinking water standard compliance data and gave no reasons for this. Two reasons can be behind this. First, these water utilities (especially the public and recently corporatized ones) might not have any (or a proper) record of their compliance with the standards and so were unable to provide data requested. It is likely that

---

[138] Personal interview on 13 May 2009.

these companies will have then difficulties when Section 29 of WSIA is fully enforced. This also applies to some private water utilities (such as SAJH and MUC). It is also possible that (some of) these companies have been reluctant to disclose their compliance record; information disclosure is not (yet) a mandatory requirement for water utilities. Some companies may regard this data as confidential and not wish to share it with an outsider.

From the information collected we can conclude that the quality of drinking water provided by water utilities in Malaysia is of a satisfactory level. One issue that clearly emerges is the urgent need to standardize the reporting and publishing of compliance data. It is remarkable that data as important as drinking water quality is not reported by the Malaysia Water Industry Guide (published by MWA) or the NWSC's water services industry performance reports (SPAN, 2010b). Drinking water quality data is a key indicator of the performance of water utilities and also of public health. In many countries this information is made public through a transparent reporting mechanism. The NWSC has to become more proactive in enforcing Section 41 and demonstrate it is capable of taking over the role previously played by the Ministry of Health in enforcing drinking water quality standards. It is not yet clear how ready the NWSC is to shoulder this responsibility.

*Readiness to comply with Section 41 of WSIA*

This section accesses the readiness of water utilities to comply with Section 41 of the WSIA. This section reads: 'the water distribution licensee shall, when supplying water to any premises, ensure that at the time of supply the quality of water supplied complies with the minimum quality standards as prescribed by the Minister' (MEWC, 2006a: 40). Sub-section (4) stipulates that contravention can render the company liable to a fine of up to RM 300,000 and/or its directors to imprisonment for a term of up to three years. This Section is anticipated to further enhance the drinking water quality standard in the country. But fully complying with its requirements may be problematic to some water utilities.

It is expected that private water utilities are in a better position to comply with the requirements of this Section. They have all the resources – financial, infrastructure, procedures, manpower – to comply with and even exceed the minimum standard. However, water utilities – private and public alike – might find it difficult to comply with Section (2), which requires water utilities to ensure 'there is no deterioration in the minimum quality standards of water which is supplied from time to time from that source or combination of sources' (MEWC, 2006a; 40). High non-revenue water (NRW) is one of the problems confronting water utilities in meeting this requirement. The most common cause of poor water quality is when muddy water intrudes into the distribution system through damaged pipes, and is then supplied to consumers. Public water departments suffer more from this problem.

Meeting this requirement will require substantial investment in asset management – pipe replacements and repairs – particularly aimed at addressing poor water quality caused by leakages. The private water utilities are better placed to invest in these works than their counterparts in the public sector. It is likely that, without sufficient investment in NRW works; water quality standards among public water utilities will further deteriorate. AWER (2011) noted that drinking water quality in Kelantan, Pahang, Sabah and Perlis was unsatisfactory. Data from surveys indicated that, at that time, financial constraints prohibited most of the companies involved from carrying out scheduled maintenance on distribution mains and undertaking active leakage control programmes, both crucial for maintaining good quality water. Attempts have been made to find solutions to these constraints. PAAB can provide flexible financial arrangements to water utilities to allow them to implement measures to improve water quality. This also requires that PAAB effectively monitors the use of these funds and ensures that investment is made in places where it is really needed for pipe replacement/repair. At the same time the NWSC is expected to set relevant and reasonable quality water standards achievement targets (for all water utilities) in accordance with the investments made. Once again this shows the importance of performance indicators as a useful tool for steering water utilities to achieve performance targets.

### 6.3.3 Information disclosure

Weil, Fung, Graham and Fagotto (2006: 155) claim that 'mandatory information disclosure by public or private institutions with a regulatory intent has become frontier of government innovation'. Mandatory information disclosure is a prerequisite for 'regulatory transparency'.

No specific studies have been carried out to ascertain environmental disclosure among water utilities in Malaysia. However, several studies have been made that analyze environmental disclosure among public listed companies in Malaysia. Those studies revealed an increasing number of companies engaging in some form of environmental reporting or disclosure practice (The Association of Chartered Certified Accountants, 2001; Yusoff, Yatim and Nasir, 2004). Other findings revealed that the majority of the companies used their annual reports for communicating environmental information. Thompson (2002) and Yusoff, Lehman and Nasir (2006) noted the growing practice of corporate environmental reporting among Malaysian companies. In their study of 40 companies listed on the Kuala Lumpur Stock Exchange, Smith, Khadilah and Ahmad (2007: 185) concluded that 'environmental disclosure is negatively associated with company financial performance'. Their finding is consistent with those of Filbeck and Gorman (2004), who also found a negative relationship between financial returns and attempts to measure environmental performance.

Under the reform, the regulation of the water supply sector in Malaysia would heavily depend on (quality) information. Without such information, it is difficult for the NWSC to undertake effective regulation. It is through this information that the NWSC can ascertain water utilities' capacity to conform with the WSIA. Such information is also needed by the NWSC, so it can determine the level of key performance indicators that it sets for each water utility. However, the information asymmetry between the NWSC and water utilities can be an impediment to effective governance. Section 29 of the WSIA exists to facilitate information gathering from water utilities (see Box 6.1). This section requires water utilities to 'furnish the NWSC with all such information relating to any matter as may require or may be prescribed' (MEWC, 2006a: 34). Furthermore, it gives the NWSC the power to appoint agents to validate the information submitted by water utilities if it deems this to be necessary. Failure to furnish information under this section carries a fine of up to RM 200,000 (MEWC, 2006a). Section 130 of the WSIA sets heavy penalties for giving false or misleading information. If proven guilty, water utilities can be fined up to RM 200,000, and their directors can be jailed for a term of up to two years or both (MEWC, 2006a). The water utilities are only required to disclose information to the regulator, and not to the general public. So there is lack of public transparency here.

---

**Box 6.1. Furnishing information under Section 29 of WSIA (2006).**

29. (1) Without prejudice to section 132, a license shall furnish the NWSC with all such information relating to any matter which:
   (a) is connected with the carrying out by the licensee of its licensed activities; or
   (b) is material to carrying out by the NWSC of any of its powers under this Act or its subsidiary legislation, as the NWSC may require or as may be prescribed.

   (2) The information required under this section shall be furnished in such form and manner, at such interval and be accompanied or supported by such explanations a supporting documents as the NWSC may require or as may be prescribed.

   (3) The information which a licensee is required to furnish to the NWSC under this section may include information which, although the information is not in possession of the licensee or would not otherwise come into possession of the licensee, is information that the licensee can reasonably be required to obtain or compile.

   (4) The NWSC may require a licensee to appoint, at the licensee's cost, an independent expert, which qualifications as may be specified by the NWSC to conduct, audit or review any of the information which a licensee is required to furnish to the NWSC under this section. The appointment and report of such an independent expert shall not relieve or derogate in any way the licensee's liability under this section.

   (5) The NWSC or its authorized officers or agents may at any time, as it deems necessary, conduct an audit on the business and activities of the licensee and the licensee shall take all necessary steps, at its own costs, to assist and facilitate the NWSC or its authorized officers or agents in conducting the audit including to grant them access to its premises and documentation and information.

   (6) A licensee who:
   (a) fails to furnish information as may be required by the NWSC under subsection (1); or
   (b) refuses to assist or facilitate, or obstructs, the NWSC, its authorized officers or agents in conducting an audit under subsection (5),
   commits an offence and shall, on conviction, be liable to a fine not exceeding two hundred thousand ringgit.

---

Given this legal requirement for information disclosure, the rest of this section analyses information disclosure in two areas: (1) compliance with EQA 1974 on sludge management, and (2) compliance with the NGDWS 2001. In addition, this section also examines how water utilities are preparing themselves to conform to Section 29 of WSIA.

## Information disclosure on sludge management

Table 6.7 shows data on the companies that do and do not record their compliance to EQA 1974 for sludge management. Notably, private water operators comply with this requirement far less than the

*Table 6.7. Recording of compliance with EQA 1974 for sludge management (n=35).*

| Ownership | no. of water utilities which record compliance to EQA 1974 | no. of water utilities which do not record compliance to EQA 1974 | No response |
|---|---|---|---|
| Public | 2 | 0 | 16 |
| Private | 3 | 6 | 8 |
| Total | 5 | 6 | 24 |

publicly owned ones. However, these figures cannot be taken as to represent the absolute picture as there were water utilities – private and public – which did not respond to the questionnaires.[139] This makes it very difficult to arrive at any conclusion about relative performance, and highlights the need for the NWSC to strengthen information reporting (and publishing) mechanisms which in turn will foster better governance.

It is interesting to investigate what prevents water utilities from recording their compliance with sludge management requirements of EQA 1974. In general, they cited three reasons for not doing so. The first reason was that the recording requirement of compliance was not mandated in their contracts (with the state governments). Second, they anticipated that DOE will eventually declassify sludge from the 'scheduled waste' list under EQA 1974, thus rendering recording requirements irrelevant. Third, there is no specific requirement under EQA 1974 for companies, including water utilities to disclosure environmental information to the public (The Association of Chartered Certified Accountants, 2002). Prior to the water sector being reformed, this information was only recorded on a voluntary basis. This is expected to change when Section 29 of the WSIA is fully enforced.

Publishing is another important part of information disclosure. Table 6.8 reveals that the majority of private water utilities (and public sector utilities) did not publish details of their compliance with sludge management requirements. Clearly they see no need to do so as this is not a legal requirement.

*Table 6.8. Publishing of compliance to EQA 1974 for sludge management (n=35).*

| Ownership | no. of water utilities which publish compliance to EQA 1974 | no. of water utilities which do not publish compliance to EQA 1974 | No response |
|---|---|---|---|
| Public | 0 | 2 | 16 |
| Private | 2 | 6 | 9 |
| Total | 2 | 8 | 25 |

---

[139] Only 11 respondents (31%) out of 35 responded to the survey (for this section).

Only private companies have so far adopted an environmental policy statement, pledge or charter. Five or 14% (of the 35 surveyed) water utilities indicated that they had such a measure in place. These had been in place between one year (GSL Water) and more than 10 years (Salcon). No public sector companies and the majority of private water utilities did not have an Environmental Management System in place and were not certain when such a plan would be implemented.

Effective regulation depends on accurate recording and reporting of high-quality and reliable information covering many aspects of water supply provisioning. This section shows that there is a lack of (high quality and reliable) information, which means that the records and reports relating to compliance with sludge management requirements are inadequate. Mandatory reporting requirements will enhance governance in the water sector.

### Information disclosure on compliance with drinking water standards

By contrast all the water utilities who responded to the surveys recorded their compliance with the NGWDS 2001 standards for drinking water quality (Table 6.9). All of the companies responding to this questionnaire were in compliance with the standards, though the compliance rate from the private sector was much higher than from the public utilities.

However, only five of the water utilities (which record their compliance) publish details of their compliance records (Table 6.10). The other six do not see the relevance of doing so, or might be reluctant to expose their compliance records to public scrutiny. Annual reports were the favoured

*Table 6.9. Recording of compliance with the NGWDS 2001 drinking water standard (n=35).*

| Ownership | no. of water utilities which record compliance NGWDS 2001 | no. of water utilities which do not record compliance to NGDWS 2001 | No response |
|---|---|---|---|
| Public | 2 | 0 | 16 |
| Private | 9 | 0 | 8 |
| Total | 11 | 0 | 24 |

*Table 6.10. Publishing of compliance to NGWDS 2001 for drinking water standard (n=35).*

| Ownership | no. of water utilities which publish compliance NGWDS 2001 | no. of water utilities which do not publish compliance to NGDWS 2001 | No response |
|---|---|---|---|
| Public | 0 | 2 | 16 |
| Private | 5 | 4 | 8 |
| Total | 5 | 6 | 24 |

method for reporting drinking water compliance, used by most of the private water companies. Public water departments do not publish annual reports.

The reform of the Malaysian water sector took place at a time of 'significant growth of information disclosure in regulatory policy' (Bennear & Olmstead, 2008). The importance of furnishing of information is made clear under Section 29 of the WSIA. This section specifies that water utilities are obliged to furnish the NWSC with 'all such information' including information on compliance with drinking water standards. However, as already explained, it does not require water utilities to disclosure their information to the public. This denies consumers' right to know and to have access to information regarding drinking water quality, sources of drinking water, detected contaminants and violations of health-based drinking water regulations. It is in stark contrast to America's Safe Drinking Water Act, which 'mandated that community drinking water systems issue annual consumer confidence report' to their customers (Bennear & Olmstead, 2008: 118). It is not clear whether the NWSC will take a similar step.

*Readiness to comply with Section 29 of the WSIA*

Section 29 of the WSIA lays down requirements and a *modus operandi* for water companies to furnish information. However, complying with this requirement would involve the water companies (especially the public ones) in a tedious process of data gathering, validation and documentation. Most of the public water companies do not yet have a proper information management system (recording, storing, retrieving, etc.) and would need to make large efforts to 'put their house in order'. This might be done as part of a corporatization exercise, which requires the utilities to list and document all the assets (and liabilities) in their possession. Private water utilities are in a better position to comply with this requirement. Most of them had already established information databases prior to being privatized. These now just need to be updated on a regular basis.

The majority of water utilities who responded to the survey (85%) were aware about the requirement for them to furnish information to the NWSC under Section 29 of the WSIA (Table 6.11). This at least shows that the essential tenets of the reform have been adequately communicated to, and understood by, the water utilities.

When asked what has been done to comply with this requirement, water utilities claimed that they will furnish the information required by NWSC. Nevertheless, some were still unclear about what information should be furnished and disclosed, and in what ways it should be furnished

Table 6.11. Awareness of information disclosure requirement under Section 29 WSIA (n=35).

| Ownership | Aware of information disclosure under Section 29 WSIA | Not aware of information disclosure under Section 29 WSIA | No response |
|---|---|---|---|
| Public | 3 | 0 | 15 |
| Private | 8 | 2 | 7 |
| Total | 11 | 2 | 22 |

and disclosed. They claim that the scope of Section 29 is too wide, as it covers all the information 'connected with the carrying out by the licensee of its licensed activities; or is material to carrying out by the NWSC of any of its powers under this Act or its subsidiary legislation' (MEWC, 2006a: 34). They thought that this section was too vague and did not define what information should be reported or published, and to whom. Several private water utilities, including PBAPP, SAJH and PNSB, believed that reporting and publishing information could improve their business image by showing that they are responsible corporate citizens. These are the same companies that have been in the forefront of integrating environmental considerations as part of their guiding principles and which want to be seen as 'environmentally sensitive' by the public and the authorities. For them, good environmental performance is vital for gaining public acceptance, especially when tariff adjustments are just around the corner. However, information disclosure can also be a source of embarrassment to some poorly-performing public water utilities. Undesired, negative, information disclosure could also inspire poorly performing water utilities to alter the way they do business (Stephan, 2002).

The importance of promoting transparent information disclosure has also attracted the attention of non-governmental organizations, such as CAWP. This NGO has called for a full disclosure of information pertaining to the water sector. It has demanded that the NWSC make all minutes of meetings, details of contracts and other documents on water available for public scrutiny (CAWP, 2007). In addition, it wants water utilities to be obliged to report and publish information on key indicators, including sludge management, water quality data, non-revenue water and service quality, and to communicate this information directly to their customers. If the amendment to WSIA is due, there is no reason why NWSC does not incorporate these important areas into the law.

It can be concluded that transparent and open information disclosure – reporting and publishing – is a prerequisite for information governance in the water sector. Information disclosure can also enable non-state actors such as civil society to have a meaningful influence on decision making concerning the reform. As such, information disclosure must go beyond a private flow of information from water utilities to the state regulator and be extended to include an exchange of information between water utilities, the state and the public at large.

## 6.4 Case study on environmental effectiveness: comparing private and public water utilities

### 6.4.1 Introduction

Section 6.3 presented an analysis of the relationship between water reform and environmental performance of water utilities. It examined whether ownership (public or private) influenced the impact of the reform on water utilities. In this section two case studies are used to further analyze the relationship of the reform on the environmental effectiveness of a private and a public water utility (PBAPP in Penang and SADA in Kedah, respectively). In the first section, the environmental effectiveness of the two utilities is examined. This is followed by an analysis of how they reacted to the reform and what changes have taken place within both water utilities as a result of the reform.

## 6.4.2 Sludge management

### Availability of on-site sludge treatment facilities

PBAPP operated fewer water treatment plants (WTPs) than SADA. SADA operated 33 WTPs while PBAPP had 10 WTPs (Table 6.12). For both companies the proportion of WTPs with on-site sludge treatment facilities was low: 30% for PBAPP and 24% for SADA. A sludge lagoon was the most common mode of treatment used by both water utilities (Figure 6.1A and B). A sludge lagoon acts as a depository for sludge before it is removed to a landfill. However, PBAPP also uses some more advanced methods, such as sludge dewatering and filtering to manage sludge at its Waterfall and Air Itam WTPs (Figure 6.1C). As required under the Environmental Quality (Scheduled Wastes) Regulations 2005 of EQA 1974, an on-site sludge lagoon can be considered as 'prescribed premises' and is therefore legally adequate for sludge treatment.

However, the main concern is with the remaining 75% of WTPs which did not have treatment facilities. Untreated sludge from these WTPs is directly discharged into rivers (Figure 6.1D). This will have adverse effects on human health and the environment. Both PBAPP and SADA indicated that land scarcity does not allow them to build sludge lagoons within the compounds of these WTPs. Acquiring land elsewhere to build sludge lagoons and contracting private services to maintain these would involve incurring additional costs. When these WTPs were built 30 years ago they were operated by public water agencies and on-site sludge treatment was not compulsory for public water agencies. The problem persisted when they took over the WTPs from state water agencies. This is because the main objective of corporatization was not to solve environmental problems, but to enhance the operational efficiency of water utilities. In addition, financial constraints have prevented water corporations from deploying green sludge treatment technologies, and they depend on state governments to provide additional land to accommodate sludge lagoons.

Both PBAPP and SADA were aware of the adverse effects of discharging raw sludge to the environment. PBAPP's Production Manager agreed that 'water utilities must not pollute the water ways' [140] but at the same time stressed that its options for complying with sludge regulations were

Table 6.12. Availability of on-site sludge treatment facilities and the type of treatment available.

| Sludge management | PBAPP | SADA | Total |
|---|---|---|---|
| # of WTPs managed | 10 | 33 | 43 |
| # of WTPs with on-site treatment facilities | 3 (30%) | 7 (24%) | 11 (25%) |
| # of WTPs without on-site treatment facilities | 7 (70%) | 25 (76%) | 32 (75%) |
| Types of treatment(s) available | sludge lagoon, dewatering, filtering | sludge lagoon | N/A |

Source: Surveys; N/A = not applicable; WTPs= water treatment plants.

[140] Personal interview on 2 Nov. 2009.

*Figure 6.1. On-site sludge management facilities. (A) Sludge lagoon (PBAPP); (B) Sludge lagoon (SADA); (C) Sludge dewatering unit (PBAPP); (D) Direct discharge of raw sludge into streams (PBAPP).*

limited. He claimed that PBAPP does not even have the capacity to handle the large quantity of sludge produced by its Sg. Dua WTP. To do so properly the company would need extra manpower, a larger area of land and to use bigger doses of chemicals.

Both PBAPP and SADA were concerned about sustainable sludge management in the future. SADA has a plan to build sludge lagoons for its remaining plants. It envisaged that 36% of its treatment plants will be equipped with sludge treatment facilities by 2012, and 40% in 2013 (SADA, 2010). However, this plan will only be implemented with financial assistance from PAAB. Similarly, PBAPP has a target of having 30% of its treatment plants equipped with sludge treatment facilities by 2012 (PBAPP, 2010). In addition, PBAPP is looking for a more advanced treatment methods, such as sludge dewatering technologies for the other plants. PBAPP is also engaged in research and development efforts to explore alternative solutions for sludge recycling, to reduce the amount of sludge and its effects on the environment.

*Sludge recycling and re-utilization: converting waste to wealth*

PBAPP was far more active than SADA in terms of sludge recycling. SADA did not see sludge recycling as a priority at the moment. Their priority is on their water supply business and it saw sludge recycling as outside this. This is in line with the decision of the state government to establish SADA as a corporatized state-owned water company whose main focus should be on running the water supply business in the state in a way can be sustained in the long run. Moreover, a representative of SADA said that they do not want to get involved in a costly sludge recycling business which has no demand for.[141]

By contrast, PBAPP's Strategic Planning Manager recognized the need to reduce the quantity of sludge disposed at a landfill by converting sludge into other usable products. The initiatives currently being undertaken include carrying out research to convert sludge into bricks.[142] PBAPP believed that sludge has the necessary components suitable for brick making, which is the red earth. PBAPP also saw the potential of promoting sludge recycling at a larger scale. PBAPP's General Manager insisted that both the government and water utilities must 'initiate the establishment of Regional Sludge Treatment Company to promote the reuse and recycling of sludge'.[143] This could lead to a coordinated sludge recycling activity, involving research and development, marketing and commercialization. He also highlighted the possibility of expanding the Indah Water Konsortium services for water sludge since 'it has proven to have the expertise and know-how and capability'.[144] This option would facilitate integration between water and sewerage services.

Currently, water utilities only undertake sludge recycling on voluntarily basis, and they would like the government to play a more active role in promoting sludge recycling. PBAPP wanted the government to make it an obligatory requirement for water utilities to allocate a portion of its revenue for research and development into sludge recycling, to financially assist the establishment of regional sludge treatment facilities and introduce economic instruments (rebate, subsidy, tax exemption, etc.) to help with the acquisition of sludge recycling equipment and technologies.

### Environmental pledge

Both PBAPP and SADA regarded environmental concern as a guiding principle in their daily operations. This is part of their bottom-line. A representative of Kedah state administration agreed that 'the environmental issue is important and water companies might go out of business if the environment is destroyed'[145]. However, he added that this awareness only seemed to be present at the policy level. At the operational level, a SADA representative lamented that they don't seriously consider environmental issues in their day to day decision making processes.[146] Moreover, the company lacks a strong written environmental pledge. The public company that SADA took over (JBA Kedah) did not and was not required to publish annual reports. When SADA came into being in 2010, a one line environmental statement was included in the company policy. It reads: 'always be aware of the environment' (SADA, 2011b).

In contrast with SADA, PBAPP was one of the water utilities which give a high priority to the environment. Its commitment on the environment was reflected in the words of its General Manager: 'PBAPP is an environmental-sensitive water utility and realizes the important

---

[141] Personal interview on 5 Aug. 2009.

[142] Personal interview on 12 Nov. 2009.

[143] Personal interview on 4 Nov. 2009.

[144] Personal interview on 4 Nov. 2009.

[145] Personal interview on 19 July 2010.

[146] Personal interview on 5 Aug. 2010.

of safeguarding the environment for long term sustainability of the water sector'.[147] PBAPP's commitment to the environment was also well documented and systematically conveyed to its employees, shareholders and public at large (in the form of annual reports). Almost every annual report it has published contains a clear and well-defined Environmental Policy, highlighting the company's commitment to safeguarding the environment (see Box 6.2). This policy re-affirms PBAPP's commitment towards developing an environmental-friendly water supply management system. This commitment was clearly translated into action where three of its facilities – the Batu Ferenggi and Waterfall plants, and Teluk Bahang Dam – received international ISO 14001 certification for environmental compliance.

*Perception of adoption of 'green tax'*

In principle, both PBAPP and SADA would welcome the introduction of a 'green tax' in the water sector. However, despite seeing this as a good move, they were not convinced that now is the right time to introduce it. The former Director of JBA Kedah thought it might take another 20-30 years before consumers would accept it.[148] Neither company expected to see the introduction of green tax in the foreseeable future, as it would not be readily accepted by water users and would place additional (financial) burdens on them.

---

**Box 6.2. PBAPP's environmental policy (PBA Holdings Berhad Annual Report 2008: 5).**

In line with its corporate objectives, PBAPP is fully committed towards protecting, preserving and conserving the environment while striving to meet all of Penang's water supply needs. Accordingly, PBAPP will:
- continually improve, update and expand its Environmental Management System which is based on International ISO14001:2004 standards;
- strive to conduct its operations in a manner that is in harmony with nature;
- reduce and/or control wastage of natural water resources and the consumption of energy and chemicals;
- conduct its business in a professional manner with an emphasis on measurable key performance indicators and results, good corporate governance and corporate social responsibility;
- prevent and avoid, as far as possible, any form of pollution by practicing proper procedures, implementing control and monitoring mechanisms, and conducting ISO14001:2004 audit practices and reviews:
- comply with all related environmental legislation and legal standards, requirements and laws set by the Malaysian Government; and
- ensure that all its personnel are fully committed towards promoting and implementing this environmental management policy in all aspects of its operations and services.

---

[147] Personal interview on 4 Nov. 2009.
[148] Personal interview on 11 May 2009.

PBAPP questioned whether there would be transparency over the use made of the tax revenues. One of the company representatives of PBAPP expressed concern as to how 'the government can assure the public that they will have cleaner rivers after they have paid the tax'.[149] Another representative of PBAPP was convinced that the prudent management of the tax would be vital for it to gain public acceptance. He proposed the establishment of a Water Catchment Agency to coordinate this initiative.[150]

If the introduction of a green tax is expected to meet with resistance among water users, PBAPP and SADA thought the government should consider internalizing environmental costs. Through this mechanism, the water companies (rather than the consumers) would be liable to pay the environmental costs. However, representatives of PBAPP warned that this mechanism might have drawbacks if the water utilities were allowed to factor in these environmental costs in their next tariff revision.[151]

### 6.4.3 Compliance with drinking water quality standards

In general, both PBAPP and SADA achieved satisfactory levels of compliance with NGWDS 2001 standards. Table 6.13 shows the percentage of water quality tests that met the NGWDS 2001 standards.

Despite intermittent violations to certain parameters, such as aluminum in drinking water, the customers of these companies generally enjoy good drinking water quality. In 2010, 96.8% of the urban and 89.7% of the rural population in Malaysia was served with clean piped water (MWA, 2011). These figures are remarkably higher than in most other developing countries. Both PBAPP and SADA have an urban-rural supply coverage that is higher than the national average

*Table 6.13. Water quality compliance standards for PBAPP and SADA.*

| Parameters | PBAPP (%) | | | SADA (%) | | |
|---|---|---|---|---|---|---|
| | 2008 | 2009 | 2010 | 2008 | 2009 | 2010 |
| Residual chlorine | 97.2 | NA | NA | NA | 98.4 | 98.5 |
| E. coli | 99.6 | NA | NA | NA | 99.9 | 99.9 |
| Residual chlorine & E. coli | 99.7 | NA | NA | NA | 99.9 | 99.9 |
| Turbidity | 98 | NA | NA | NA | 95.6 | 96 |
| Aluminum | 89.9 | NA | NA | NA | 92.1 | 93 |

Source: PBAPP and SADA documents; NA = data not available.

---

[149] Personal interview on 4 Nov. 2009.

[150] Personal interview on 4 Nov. 2009.

[151] Personal interview on 2 and 12 Nov. 2009.

(MWA, 2011). Both companies consistently strive to continuously produce and supply quality water to its consumers. Apart from pipe replacement and repairing, both have also focused on laboratory accreditation. In 2008, PBAPP received MS ISO/IEC 17205 standard[152] accreditation for laboratory testing and calibration. This accreditation ensures that PBAPP' labs are up to the mark in supporting its core objective of maintaining good water quality standards.

SADA's laboratory accreditation was mainly conducted by its concessionaires: there is no record of any of SADA's laboratories being accredited. In 2008, Taliworks's Sungai Baru[153] laboratory received MS ISO/IEC 17205 standard accreditation from the Malaysian Department of Standards. During a field visit to Padang Saga water treatment plant[154] (run by a concessionaire) it was noticed that the laboratory was better equipped and administered than those of SADA. The plant controller confirmed that this laboratory was administered by one full time chemist and capable of conducting tests on a range of water quality parameters such as pH, turbidity, chlorine and fluoride.[155]

Overall, PBAPP and SADA are confident that the reform and PAAB's financial approach will facilitate them to meet the drinking water quality standards required under Section 41 of the WSIA. As such, both believe that they can maintain and achieve (high) water quality standards, as reflected through their commitment to meeting NWSC standards in the coming years (Table 6.14).

*Table 6.14. Water quality targets (2011-2013) – based on % of water quality tests that meet the NGDWS 2001 standards.*

| Parameters | PBAPP (%) | | | SADA (%) | | |
|---|---|---|---|---|---|---|
| | 2011 | 2012 | 2013 | 2011 | 2012 | 2013 |
| Residual chlorine | 98.5 | 98.5 | NA | 98.5 | 98.5 | 98.5 |
| E. coli | 99.9 | 99.9 | NA | 99.9 | 99.9 | 99.8 |
| Residual chlorine & E. coli | 99.9 | 99.9 | NA | 99.9 | 99.9 | 99.9 |
| Turbidity | 97 | 98 | NA | 97 | 98 | 98.5 |
| Aluminum | 94 | 95 | NA | 94 | 95 | 95.5 |

Source: PBAPP and SADA document; NA = data not available.

[152] Awarded by Malaysian Department of Standards.
[153] Taliworks Corporation manages the water supply on the island of Langkawi on behalf of the state government/SADA.
[154] Managed by Taliworks Corporation.
[155] Personal interview on 9 Aug. 2010.

## 6.4.4 Information disclosure

PBAPP appeared to fare better than SADA in terms of recording and reporting compliance data for sludge management and drinking water quality standards (Table 6.15). PBAPP is one of the most efficient (and profitable) water utilities and this has given them the advantage of being able to acquire digital technology for its data management (including that on compliance to sludge management and water quality). PBAPP is taking active steps to get all its plants recording their compliance with EQA 1974 and NGWDS 2001 by 2011.

Public water utilities, such as JBA Kedah/SADA do not have such a good record in making information available for public consumption. Information reporting, if done at all, is done mainly for internal use. By contrast, private water operators, such as PBAPP, communicate their compliance on both EQA 1974 and NGWDS 2001, mainly through their annual reports presented to the Kuala Lumpur Stock Exchange and their shareholders. Publication of an annual report is a mandatory requirement for publicly-listed companies. PBAPP has also used leaflets and its website to publish environmental information. However, the level of detail of information was quite limited. A quick look at PBAPP's annual report reveals that no reference whatsoever is made to any specific parameters relating to compliance to NGWDS 2001 (PBAHB, 2008: 28).

Not only having a detailed environmental policy in place, PBAPP also realizes the importance of communicating its quality policy to the general public, shareholders and employees. PBAPP regularly communicates its quality policy through various mediums: its annual reports, internal communications, leaflets, brochures and its website. Box 6.3 documents PBAPP's commitment towards achieving a high quality service in supplying quality drinking water.

This is insignificant contrast to when JBA Kedah was the responsible for water supply, as environmental concerns were never explicitly documented or communicated to the public. If

*Table 6.15. Information disclosure for PBAPP and SADA.*

| What | EQA 1974 | | NGWDS 2001 | |
|---|---|---|---|---|
| | **PBAPP** | **SADA** | **PBAPP** | **SADA** |
| Recording | Yes | No | Yes | Yes |
| Reporting/publishing | Yes | No | Yes | No |
| Method of recording | Digital | Non-digital | Digital | Non-digital |
| Method of reporting/publishing | Annual report, leaflet | No reporting | Annual report | No reporting |
| Medium of recording | Individual plant | Overall compliance | Individual plant | Overall compliance |
| Medium of reporting/publishing | Individual plant | NA[1] | Individual plant | NA |

[1] NA = data not available.

> **Box 6.3. PBAPP's quality policy (PBAHB Annual Report, 2008: 4).**
>
> In line with its corporate objective, PBAPP is fully committed towards continual improvement as it strives to provide high quality services and products that will satisfy and delight customers.
>
> Accordingly, PBAPP will:
> - continually to improve and update its Quality Management System, which is based on international ISO9001:2000 standards;
> - sustain a corporate culture driven by continual improvement by promoting and encouraging innovation, teamwork, diligence and creativity, as well as a proactive approach to water supply services;
> - provide the best possible training opportunities to encourage its employees to continuously upgrade their competence levels, knowledge and skills;
> - uphold its reputation as a model water supply organization in Malaysia;
> - ensure the protection, preservation and conservation of the environment;
> - provide a safe and healthy working environment for all its personnel; and
> - ensure that all its personnel are fully committed towards promoting and implementing this quality management policy in all aspects of its operations and services.

recorded at all they were mostly meant for internal use, and produced and communicated in print form rather than using new media such as Internet. Furthermore, like other public departments, publishing an annual report was never a practice or norm within JBA Kedah. This practice started to change when SADA was established in January 2010. While the reform has not fully turned SADA into a corporate entity, it has to a certain extent forced SADA to pay greater attention to quality improvement within its core business, water supply. Its environmental concerns, especially in the area of the quality of services, were made more explicit. On-line media has been added to printed media as means of communication. Box 6.4 sets out SADA's new commitment towards quality water supply services (SADA, 2011a).

> **Box 6.4. SADA's quality policy (www.sada.com.my).**
>
> SADA is committed to attaining the company's vision and mission through an integrated quality management approach, continuous improvement, a high level of work professionalism and compliance with laws and the company's policy in order to provide the best services to all customers at all times.

Section 29 of the WSIA requires all water utilities[156] to furnish information to the NWSC. This information has to be furnished in 'such a form or manner, at such intervals and be accompanied or supplemented by such explanations and supporting documents as the NWSC may require or as may be prescribed' (MEWC, 2006a: 34).

Both PBAPP and SADA are aware of this requirement and do not anticipate any problems in complying with it. Both have been supplying information to the NWSC on a yearly basis on almost every aspect of water operation – key performance indicators, financial data, customer services and technical data. However, PBAPP felt that the gathering of information could be enhanced by using a standardized reporting format. This would ensure that every party – NWSC, MEWC and state water regulators – is given and receives the same information in the same format. The PBAPP proposed that the NWSC should take control of coordinating information gathering (from water utilities) and minimize unnecessary duplication of information.

PBAPP and SADA acknowledged the importance of the requirements to furnish information in enabling NWSC to perform their regulation duties effectively. Both companies agreed that information was the most single important factor needed to regulate water utilities. Both are aware that, without (reliable) information, it was difficult for NWSC to propose accurate, measurable and achievable key performance indicators for water utilities. Elsewhere in the world, such information has been widely used as a tool for benchmarking amongst water utilities. For instance in the UK, information serves as an essential component for comparative regulation. PBAPP and SADA felt that benchmarking could steer under-performing water utilities to emulate the best practices of the best-performed ones. They also thought that this information allows the public the avenue to judge the performance of water utilities, and to decide whether the tariff they are paying is commensurate with the quality of services they receive. Thus information will drive the (under-performing) water utilities to improve their operations and services, move towards meeting key performance indicators imposed by the NWSC and secure public trust. This is in line with objective of the reform in promoting transparency in the sector through information disclosure.

### 6.4.5 Conclusions of the case study

Overall, it is not possible to establish a clear relationship between the water sector reform and any changes in the environmental effectiveness of water utilities. The two in-depth case studies further re-affirm that the performance of both private (PBAPP) and public (SADA) water utilities on three indicators were dependent on or subject to conditions that existed before the water sector was reformed. Sludge management, water quality compliance and information disclosure, are all fairly time-consuming undertakings and it is unlikely that the reform could contributed to significant changes in such a short period of time.

Still, it can be concluded that it is important for environmental effectiveness to continue in the reform process. The two case studies have indicated the areas in which water utilities are doing reasonably well and where there are pressures for them to improve (Table 6.16). Water utilities can be expected to work hard to improve their compliance with sludge management and information

---

[156] Under this section, they are referred to as a licensee – a person whom was granted a license by NWSC to operate water (and sewerage) system.

*Table 6.16. Summary of environmental effectiveness for PBAPP and SADA.*

| Indicators | PBAPP | SADA |
|---|---|---|
| Sludge management (EQA 1974) | | |
| • Availability of on-site treatment facilities | Low | Low |
| • Adoption of environmental-friendly treatment facilities | Yes[1] | Yes[1] |
| • Sludge recycling initiative | Yes | No |
| • Written environmental policy/pledge | Yes | Yes |
| • Perception of adoption of green tax | Good idea. Implementation might be problematic | |
| Drinking water standard | | |
| • Compliance to NGDWS 2001 | Complied | Complied |
| • Readiness complying with Section 41 of WSIA | Ready | Ready |
| Information disclosure | | |
| • Recording | Yes | Yes |
| • Publishing/communicating | Yes | No |
| • Having written Environmental Policy/Charter | Yes | No |
| • Readiness to comply with Section 29 of WSIA | Ready | Ready |

[1] Adoption time not indicated.

disclosure, and have to place less emphasis on water quality (which is already at acceptable levels). The analysis indicates how the reform facilitates and pressures water utilities to make improvements in these areas. Mechanisms and approaches, such as the financial arrangement (through PAAB for sludge management) and regulation (WSIA for water quality and information disclosure) have already been established to drive this process forward. Nevertheless, it should be noted that these mechanisms and approaches have not yet produced significant results.

## 6.5 Conclusions

This section has analyzed the effects of the water reform on the performance of water utilities in relation to three environmental indicators: sludge management, drinking water quality compliance and information disclosure. In this analysis, a comparison was made between public (state water departments and corporatized water utilities) and private water utilities, and this was furthered refined in the two case studies. Three conclusions can be drawn from this chapter.

The first one is related to the contribution of the reform particularly in enhancing the implementation of the performance indicators for the Malaysia water sector. As indicated, none of these indicators are used to measure the environmental performance of water utilities at the moment (MWA, 2011; SPAN, 2010a). This, to a certain extent, is indicative of water utilities in Malaysia being generally less environmentally conscious than their counterparts in developed countries (Lee, 2010). While the indicators proposed here are not exhaustive (for measuring environmental performance), they do at least lay the basis for recognition of the inextricable link between water supply and the environment/health. Other indicators might well be included later. With the reform already well in place, in particular the institutional reform (NWSC) and legal framework (WSIA and SPANA), these indicators could be easily adopted, systematically monitored and enforced. This would strengthen the extent to which the reform actually helps to protect the environment/health, rather than being a mechanism for 'economizing on the environment' (Barry, 2005).

In the second place, we can conclude that private water utilities seem more capable of attaining higher compliance levels with sludge management and information disclosure requirements than their counterparts in the public sector. Both sectors have a good compliance level to with the drinking water quality standards contained in NGDWS 2001. Even though, it is too early to equate changes in environmental effectiveness with the reform, its findings suggest that the reform is able to further trigger improvements in the water sector. The reform gives NWSC the opportunity to intervene in areas where there is a low level of compliance or of readiness to comply (i.e. sludge management and information disclosure). One available option is to encourage the public to put pressure on poorly performing companies. Another is to provide incentives and rewards to the good performers (Lee, 2010). In exercising this option, fairness is of paramount importance, thus (at all costs) NWSC must avoid 'cherry-picking'. The reform must be able to recognize disparities – in financial and technological resources – between private and public water utilities, and how these can affect companies' ability to comply with environmental requirements.

A third and last conclusion relates to information. The information (flow) transcends the information asymmetry between private-public water utilities and regulators-regulatees trajectory (as it was before). It also goes beyond the importance of (high quality and reliable) information for effective regulation and includes two other important aspects. The first is that disclosed information must be integrated into the decision making processes of the water utilities and regulators (Weil *et al.*, 2006: 159). It also implies that (environmental) information needs to be embedded in the formulation of regulatory tools (by NWSC), business decisions made by water utilities and the behaviour of water users (so they adopt a greener consumption pattern). Overall, it calls for strengthening the link between information holders/disclosers and information users (Lee, 2010). The second point is that information should no longer be a privileged resource, confined only to water utilities and the state regulator. The public, consumers and NGOs also have a right to access to this information, which must be upheld as it will lead to greater transparency. Water utilities must be held accountable, not only to the regulator and their shareholders but, most importantly, to those who pay them, their customers. In this respect, Sections 29 and 132 of the WSIA will have to be enforced in due course. The Water Forum must be allowed to take its proper place as the platform to promote public scrutiny of the water sector.

# Chapter 7.
# Conclusions

## 7.1 Introduction

For decades the water supply sector in Malaysia has been confronted with four main problems: operational inefficiency; governance and regulatory ineffectiveness; budgetary constraints; and environmental ineffectiveness. As highlighted in Chapter 2, the operational inefficiency in the sector stemmed from the inability of water utilities to reduce the high level of non-revenue water (in some states the level of water losses were as high as 50%) to below cost recovery levels, leading to water utilities having insufficient revenue to sustain and expand their operations.

The system of governance in the water sector was ineffective, since responsibility for formulating policy, regulating water provision and providing water were in the hands of one single body: the state governments. The constitution (which gave the state governments sole jurisdiction in this area) limited the extent to which the Federal Government could intervene in this matter. Furthermore, the involvement of state governments in joint-venture water businesses with private actors served to further erode their role as guardians of citizens' rights to water. In Selangor and Johor, for instance, the state governments were criticized for not safeguarding the interests of water users with regard to water tariffs. State governments were also unable to effectively regulate the behaviour of water utilities, which often had political ties with the state administration.

A high level of dependence on limited public funding restricted the ability of the water sector to invest in developing the water infrastructure. The budgets allocated under the five-year Malaysia Plan were far short of the amount required by the state governments for water infrastructure. This problem resulted in delays in investment in several crucial areas, such as tackling water losses, extending water coverage and modernizing the information and IT systems. Moreover, (fragmented) privatization did not bring tangible benefits to the state governments, especially when profitable (treatment) operations were given to private operators with exclusive rights for tariff revisions. For political reasons, the state governments were reluctant to grant tariff increases to water operators, which eventually resulted in the state governments having to pay huge sums of compensation to private water operators. In the end, the water users ended up as the victims of this struggle, by paying higher tariffs.

Finally, the water sector came under pressure to find durable solutions to its poor environmental performance in terms of sludge management and information disclosure, and – to a lesser extent – water quality. Rapid urbanization and industrialization increased water demand, which in turn led to a more water treatment sludge being generated. Without proper treatment and disposal, the direct discharge of, the high toxic, raw sludge posed serious environmental problems, and establishing sustainable sludge management would have been a financially burden to water utilities (already facing revenue problems). Lack of government incentives and of any commercial demand meant that sludge re-cycling and reuse was not an attractive proposition for water utilities and the companies lacked the finance to invest in environmental-friendly production technologies. With respect to information disclosure it proved that not only acquiring information about the water sector was a difficult task, the information that was available suffered from validity problems. This was caused by the absence of any clear policy and of a dedicated body for information management. Moreover, public disclosure of information was prevented by laws. The information asymmetry (between the private water sector, the regulator and civil society) combined with the scarcity and poor quality of data prevented the regulator from fulfilling its task effectively.

The desire to address these problems motivated the government of Malaysia to initiate the reform of the water sector in 2004. The primary objective of reform was to improve the efficiency and effectiveness of the water sector in the long term. This research assesses the extent to which the reform has met its intended objectives and the means through which it has done so (or not). Three research questions guided this research:

1. How can we understand and explain the policy process of the water sector reform?
2. To what extent have the outputs of the reform contributed to the realization of the reform's objectives?
3. To what extent has water sector reform improved the operational efficiency and environmental effectiveness of water utilities?

Section 7.2 syntheses the main answers to these research questions. Section 7.3 highlights the contribution of this study to the wider literature on water sector reforms. Sections 7.4 and 7.5, respectively, reflect on the theory and methodology used in this study, Lastly, in sections 7.6 and 7.7, areas for future research and policy recommendations are formulated.

## 7.2 Synthesis of the main findings

This study was conducted to assess the (interim) effects of the reform on the water sector. It does so by comparing the performance of water utilities in the pre-reform period (2004-2006) and in the post-reform period (2007-2009). The reform of the Malaysian water sector occurred against the background of an (on-going) discourse about the global water sector and how to improve effectiveness and efficiency. Here, I synthesize the main findings to the study's research questions.

### 7.2.1 Understanding and explaining the policy process

Three conclusions can be made regarding this question. First, the policy process of reforming the Malaysian water sector was representative of the global trend to centralize water management within the public – rather than the private – domain. The centralization of water management within the public domain facilitated good governance, since it institutionally separated (the previously overlapping) responsibilities for policy formulation, regulation and service provision. The reform was also characterized by a strong drive from the government towards corporatization, evidenced by the formation of a number of water corporations and the decreasing role of private water operators. This approach was favoured as it consolidated the strengths of both public and private approaches to water management. This strategy acknowledged the necessity of the public domain retaining control over water, and of including a private culture in water operations to address inefficiency and ineffectiveness. The introduction of a private sector working culture into public water authorities required effective regulation (undertaken by an independent body). This practice of placing regulatory oversight in public hands is increasingly common in the water sector worldwide (see Kessides, 2004; Nyarko, 2007). Schwartz (2008) refers to this approach to managing water as 'new public management' and Rouse (2007) uses the term 'corporatized-public water companies model'.

Second, and in line with the first conclusion, this research provided evidence of the dominance of the federal and state governments in the water sector reform process. This was strongly influenced by the failures of privatization in many parts of the world (see Casarin *et al.*, 2007; Araral, 2009; Ahlers, 2010). The private water operators in Malaysia did not possess the necessary resources to influence the discourse enough to challenge the state actors. Hence, private water operators (such as AUIB) were disappointed when corporatization limited their participation in larger water ventures, and their roles were reduced just a small part of the country's water operations, such as operating and maintaining the treatment plants. The dilution of private powers (mostly private water operators, but also international donors) is also noticeable in the new ways in which the water sector is financed after the reform. PAAB now provides an alternative to purely private water financing, thus undermining the previously powerful position of private actors in this area. Nevertheless, the government also acknowledges that the dominant role of the state actors can have detrimental effects to the sector. But rather than enlarging the role of the for-profit private sector, more possibilities were opened up for civil society involvement. Even though its role has not yet been significant, their presence has at least been acknowledged (Kamat, 2004). The establishment of the Water Forum could further enhance civil society involvement in post-reform decision making. But, as observed during the interviews, this body can only serve as an effective public platform if it is allowed to function free from influence from the water companies; a condition which the civil society organizations urge the central regulator to uphold.

Third and lastly, this kind of policy process is usually mediated by a set of established formal and informal rules and procedures (as summarized in Chapter 3). In this reform process a number of new and unusual rules and procedures were created. The direct participation (in weekly Cabinet meetings and in three special meetings with all stakeholders) of the Prime Minister and his Deputy in the reform process is rather unique. The involvement of both men was crucial in resolving conflicts between federal and state agencies and in formulating and achieving a common goal. This approach could be replicated in the policy formulation processes in other sectors; even though there is no indication of it being applied elsewhere. A further unprecedented change was the increased involvement of the public in the reform process. The disclosure of public documents, the WSIA and SPANA, to the public through the Internet before both documents were tabled to the Parliament is an example of this, as was the use of the public hearings to solicit public feedback on the proposals. Over 50 public hearing sessions were held between early 2005 and mid-2006, allowing all stakeholders (including members of the opposition parties) the opportunity to comment on the two sets of legal documents. Later, the same approach was adopted by the central regulator to solicit public feedback before finalizing several of its regulations and rules for the water sector. Public involvement in the water sector reform was established in order to secure public 'endorsement' and support for a rather radical policy change, where power balances were clearly shifting.

### 7.2.2 The outputs of the reform and how they contributed to achieving the reform's objectives

From the evaluation of the outputs of the water sector reform, we can draw four conclusions about the success of the reform in achieving its objectives. These relate to regulation, water resource management, financing and operational issues.

The failures of regulation prior to reform clearly demonstrated the need to establish an effective regulatory oversight, a feature also highlighted by many scholars (such as Rouse, 2007; Holland, 2005). Araral (2010) concluded that establishing an effective and independent regulatory body in the water sector is always difficult in practice, and only 21% (out of the 122 developing countries he surveyed) has a well-functioning independent regulatory agency. One of the main objectives of the reform was to detach the regulatory responsibility from the state governments (which had too many, conflicting, water tasks, and were too closely related to the water companies) and to place it with an independent body, an approach based on the OFWAT model in the UK. Basically this was achieved by relocating many of the water regulatory tasks (on water supply and services) within one independent institution at the federal level, the NWSC.

But truly free and independent regulation will probably never exist in the Malaysia water sector, and arguably in no country. In Malaysia, the new regulators are appointed by politicians (i.e. the Minister) and their actions are bound to be subjected to the discretion of the Minister and the Ministry. Several interviewees questioned the impartiality of the NWSC. According to the Regulatory Director, in contemporary Malaysia 'the kind of regulating independency needed must fall within our own mould' (personal interview 15 Sept. 2009). Franceys and Gerlach (2011) noticed similar tendencies in Jakarta (Indonesia), where the water regulator is filled with local politicians. By the same token, political ties can also help the regulator to carry out certain actions. The NWSC's political ties and support have helped it to successfully complete the corporatization of five state water departments and implement tariff revisions in four states.

This shows that the reform has (at least to some extent) established effective regulation in the Malaysian water sector. The successful tariff revisions in four states show that centralizing the regulatory function has also reduced the influence of state politicians on tariff setting. But the new regulatory institution is not yet fully stable in terms of its human resources, information availability, expertise and relations with state water departments and water corporations. Hence, Pigeon (2012) is correct in observing the difficulty of 'de-politicizing' tariff decision making in a situation where the regulatory oversight is weak and lacks expertise.

The second conclusion relates to the remaining role of the state regulators in managing water resources. Following the reform, state water resource regulators were established to safeguard water resources: a crucial complement to the role of the central regulator on water supply and services. We can draw two clear conclusions about the role of these state water resources regulators. First, these bodies have not been effective in safeguarding water resources from unplanned economic activities, such as logging or sand dredging. They have been criticized by civil society organizations and federal agencies for failing to effectively enforce environmental laws pertaining to water resources. Second, a lack of political will (from the state government) and potentially significant economic losses are main factors behind the poor track record of these bodies. It is essential that the federal government assists the state water resource management organizations to fulfil their tasks and provides financial support to offset any economic losses stemming from strict environmental law enforcement.

With respect to financing, the outputs of the reform have better met the set objectives. PAAB has been the key institution for addressing the over-dependency of state governments on limited public budgets to develop water infrastructure, enabling relatively cheap access to foreign capital. State governments are confident that PAAB will be able to provide the required budgets to develop water infrastructures. Six state governments have already agreed to subscribe to the PAAB financial mechanism and negotiations with other state governments had reached their final stage by 2012. Through this approach, water utilities will benefit from relatively cheap loans obtained through PAAB, which will help in cushioning tariff increases that might otherwise have been caused by private borrowing. Nevertheless, several uncertainties remain about PAAB's approach. Civil society organizations, such as CAWP, and some water utilities are concerned whether this approach will really be beneficial, since PAAB will be exposed to foreign borrowing risks. The utilities are afraid that PAAB might impose higher lease rentals especially on the richer private operators. PAAB might also use federal government backing (for cheap loans) to become a money-making venture. Private operators such as SAJH have already been charged higher lease rentals (6% annually) than public water departments (3-4%). NWSC is responsible for ensuring that PAAB's mandate of providing sustainable and affordable financial mechanisms remains unchanged.

The fourth and last conclusion relates to water operations. The establishment of state water corporations has helped inject efficiency in the water sector, without the state fully losing control over water operations. Corporatization introduced efficiencies in four ways:
1. by managing water provisioning as a private business venture under capable professional management;
2. by bringing water corporations under stringent and systematic regulation from an independent central regulator;
3. by reducing the influence of state politicians on water operations (i.e. in tariff decision making and in the operation of water corporations); and
4. better access to (cheaper) finances (through PAAB) which has facilitated improvements in the efficiency of water supply and the performance of water corporations.

Even though the existence of water corporations has threatened the ability of the private water operators to expand their operations, its contribution in ensuring that water will remain in public hands is well received by civil society organizations. It can be concluded that effective regulation and a viable financial mechanism are both prerequisites for water corporations to function effectively.

In conclusion, the analysis of the effectiveness of the outputs of the reform shows mixed results. Overall, the central regulator, water corporations and the financier have proved themselves to be important institutions that meet the objectives set for them. By contrast, the state water resources regulators have not (yet) shown themselves to have the capacity to meet the objective of safeguarding water resources.

### 7.2.3 The operational efficiency and the environmental effectiveness of water utilities

For the final research question, the outcomes of the reform were analysed by comparing the operational efficiency and environmental effectiveness of the water utilities before and after the reform and by comparing the performance of public and private operators. Three conclusions can be drawn regarding this question.

First, it proved to be very difficult to link the performance of water utilities (on either set of the indicators) to the reform. Overall, there was no clear evidence to link operational efficiency and environmental effectiveness with the reform. Although some indicators, such as non-revenue water and unit production costs, have improved since the reform, it is probable that these improvements were due to other factors. The decrease in levels of non- revenue water started before the reform and was the result of the gradual and on-going implementation of initiatives by water utilities to reduce water losses. It can also be concluded that the reform had little effect on motivating water utilities to deploy environmentally-friendly sludge treatment technology or building sludge lagoons. Similar conclusions can be drawn from the comparison of the performance of public and private water utilities. Even though the research revealed that private water utilities seemed more competent at managing water losses and in their collection efficiency, the in-depth case studies suggest that the difference in performance between these two models of water utility was not due to the reform. Overall, the differences in the performance of water utilities (both prior to and after the reform, and between public and private water utilities) are more dependent upon external factors than to reform-related factors. But this does not imply that water sector reform has not had any influence on performance. Lack of (reliable) data on the different indicators and the short time interval between the reform and this research make it difficult to draw firm conclusions on the effect of the reform on operational efficiency and environmental effectiveness.

The second conclusion relates to the use of performance indicators. In developed countries (i.e. the Netherlands, the UK), water utilities use a set of standardized performance indicators to measure environmental performance. This practice is not yet well-established in the Malaysian water sector. However, the reform has contributed to increased utilization of standardized performance indicators (on operational efficiency and environmental effectiveness). Yet, as this research highlights, there is still a lack of (reliable) data and the set of performance indicators used is too limited. While some indicators, such as non-revenue water, collection efficiency and compliance with drinking water standards are commonly used, (reliable) time series on all water corporations do not yet exist. Other relevant indicators will need to be added as the sector develops. Sludge management and information disclosure are two important indicators that are currently missing from the NWSC indicator lists and also from the Malaysia Water Industry Guide published by the Malaysia Water Association.

A third and related conclusion is on information disclosure. The water sector needs to further improve its provision of (reliable and quality) information, to both water regulators and to the public. Reliable and quality information is not only crucial to facilitate the NWSC to measure the performance of water utilities, but also for civil society and consumers to judge water corporations. The information asymmetry (or gap) between the information-rich (private) water utilities and NWSC and the public has to be closed. This is in line with Lee (2010) who called for stronger links between information holders/disclosers (water utilities) and users (the regulator and the public).

The NWSC has recently devoted substantial time and resources in developing an information management system, to collect, record, retrieve, validate, and control the quality of data. Section 29 of the Water Services Industry Act sets standards here, making information disclosure mandatory, and giving the regulator the power to check the information submitted by water utilities. It is also stipulates that is a crime to provide false information. All this helps to strengthen the position of the regulator, but it is unclear what mechanisms exist to provide civil society with sufficient and reliable information.

## 7.3 Contribution to the wider literature on water sector reform

This research was conducted in the context of Target 10 of the United Nations' Millennium Development Goals – to halve the proportion of people without access to clean water supply and adequate sanitation by 2015 (Asian Development Bank, 2006). Meeting that target demands that water suppliers in developing and transitional countries make serious investments in human resources and technical development. The Asian Development Bank estimates that US$ 8 billion annually is required for countries in Asia and the Pacific to meet that target. The Bank also highlights the growing awareness that the barriers to achieving this target are frequently political and institutional, rather than economic or technological. Water sector reform can help countries to move towards this target.

In this context, the lessons from Malaysia's experience are of potential relevance to other (developing and under-developed) countries. This section explores how the Malaysian water sector reform contributes to current discourses about ways to improve the water sector in such countries.

### 7.3.1 Public-private partnership

Even though the failure of water privatization has been cited as one of the factors that drive countries to implement water sector reform, the private sector often continues to play an important role (see Cook & Uchida, 2008; Ehrhardt & Janson, 2010). But, when water management is turned into a business venture, the behaviour of the water corporations must be properly regulated, yet many countries have undertaken privatization without putting sound regulatory oversights in place. Countries can fruitfully tap into the strengths of the private sector (with respect to its working culture, operational efficiency and financial resources) to improve the efficiency and effectiveness of the public water sector. The water corporation does that. Malaysia's water sector reform shows how a monopolistic water sector can successfully marry the public sector (in policy formulation, regulation and water rights) and the private sector (in working culture and financial resources). This research re-iterates the importance of not just focusing on increasing private sector involvement in water management, but their contributions can only be enhanced through effective regulation. Scholars such as McDonald (2012: 9) regard this as 'a new counter-narrative to the liberal neoliberal ideology of market-based service delivery solutions'.

### 7.3.2 Effective and independent regulation

This research has highlighted the importance of effective and independent regulation of the water sector, irrespective of whether it is public, private or corporatized. Independent regulation means that the sector is (relatively) free from political interference, while effective regulation implies the need for sufficient and reliable information and strong regulatory enforcement.

In many parts of the world, especially those with unstable and/or undemocratic political systems, the requirement for independent regulation is difficult to fulfill. As this research has re-affirmed, a truly independent regulatory oversight is difficult to achieve in developing or transitional countries and it may even hamper water sector reforms, if they rely on political support. Hence, the most appropriate form and degree of independent regulation will depend on the existing local socio-economic and political conditions, as this study shows. The form and degree chosen in Malaysia might well be relevant for other developing Asian countries as these countries, to some extent, share similar local conditions. We can draw a similar general conclusion on effective regulation in many developing Asian countries, which face poor availability of (reliable) information for the regulator and weak enforcement.

### 7.3.3 Informational governance

Third and lastly, this research contributes to the general debate about informational governance. The Information Age makes it possible for all citizens to have wider access to information, through the Internet, satellites, interactive television and mobile phones (Mol, 2008). But this research showed that the water sector in Malaysia seems to have only partially stepped into the Information Age, with limited information generation, recording, validation, utilization and dissemination. Malaysia (and other transitional and developing countries) still have relatively weak informational governance, due to a combination of economic and socio-political factors. Information collection and storage is limited and remains largely in the possession of the government and private enterprises. Public disclosure of information is often prevented by laws (i.e. the Official Secrets Act 1972 and Banking and Financial Institutions Act 1989 for financial data in Malaysia) that limit public access to information, which is counter-productive to promoting informational governance. Informational governance needs to be promoted in developing countries, through making information more available and using it more in decision making processes and public debates, thereby creating accountability (Xia, 2010). The Malaysian water sector reform shows some first steps in enhancing information disclosure in order to promote good governance, enhance democratic policy-making processes and further involve civil society in exerting countervailing power. This reflects Mol's findings (2009) about the environmental sector in China.

## 7.4 Reflections on theory

This research was approached from two theoretical frameworks: the policy arrangement approach (PAA) – applied to understand and explain the policy process of the water sector reform – and the policy evaluation approach (PEA) – used to analyze the outcomes of the reform on the operational efficiency and environmental effectiveness of the water utilities. Two conclusions can be drawn about their value and limitations (both separately and in combination with each other).

First, it can be concluded that the two approaches complemented each other as theoretical frameworks. For instance, PAA was of great relevance in understanding the general policy process in various social domains, not limited to the water sector (e.g. Arts & Van Tatenhove, 2005; Arts & Goverde, 2006; Liefferink, 2006). But the application of PAA does not help to evaluate the impacts of the policy process on target groups. Thus, this research relied on PEA to investigate the impacts of the water sector reform on the operational efficiency and environmental effectiveness of water utilities.

PEA allowed a refinement of the outcome analysis, by looking at the performance of water utilities on four indicators for the operational efficiency – non-revenue water, collection efficiency, unit production cost, customer service complaint – and three for environmental effectiveness – sludge management, water quality and information disclosure. Nevertheless, as indicated in Chapters 5 and 6, PEA only partially explains the causality between the reform and the performance of water utilities. The differences in the performance of water utilities were also influenced by external factors – which existed prior to the reform, or before the water utilities were privatized or corporatized. The four dimensions of the PAA came in useful here in investigating and explaining these factors. Private water operators were generally better at managing water losses and collection efficiency, because they have more access to financial resources and superior technological know-how. Equally it showed that despite having access to legal resources and water rights the state water resource regulators were not effective. The research showed that they are heavily depended on the political will (of state governments) and financial resources from the federal government.

Using PAA and PEA in combination was very effective. It allowed the analysis to go beyond mere 'understanding' and beyond the linear causality of 'impact evaluation'. A combination of the two theoretical approaches could also be used to evaluate the performance of other utility sectors such telecommunications, electricity or gas.

Second, the PEA, as developed and used here, is a context-specific model (whereas PAA has a more general validity). The PEA model has been developed and widely applied within developed countries, where information gathering, recording, validating, utilization, dissemination are extensive and there is great respect for the freedom of information. In such circumstances the PEA model works quite well. In information-poor environments it is more problematic to apply the PEA model, due to information shortages/scarcity and unreliable information, which make the assessment of outcome and impact effectiveness a near impossibility (Mol, 2008). This raises doubts about its relevance in situations where data scarcity and limited information disclosure make it difficult to evaluate certain indicators. This was partly experienced in this research, particularly when seeking to analyze the quantitative indicators of the reform (i.e. non-revenue water, collection efficiency, etc.). In data-poor environments the use of expert judgments might be a better way to assess performance than using abstract (and incomplete) quantifiable datasets.

## 7.5 Reflections on research methodology

This study has also raised several methodological questions, three of which are worth reflecting upon here: the time scope for policy evaluation, data availability, and the use of case study research.

### 7.5.1 The time scope for policy evaluation

Reform of the Malaysian water sector reform began in 2004 and the reform process entered the implementation stage in 2007 (with the establishment of laws and institutions). The reform is far from finished or stabilized and is very much an on-going process. In this respect, this research represents an *ex nunc* evaluation of the impacts of the reform. As specified in Chapter 2, this research evaluates the 'interim' impacts of the reform on the performance of water utilities in the years 2007-2009.

Performing an *ex nunc* evaluation has both advantages and disadvantages. The clear advantage is that it can identify areas for improvement and policy modifications, which – when implemented – can increase the likeliness of meeting the intended objectives of the reform. For instance, several key policy areas were identified where there is a need for further modification in order to reach the objectives of the reform. These include bringing water resource management within the scope of the reform, establishing a single body to handle water issues, the formation of regional sludge companies to spearhead research and development in sludge recycling, and the internalizing of environmental costs in water tariffs. If the similar evaluation were conducted after the reform was fully stabilized, it would be harder to include such policy recommendations in the on-going reform process.

A second advantage of a timely evaluation is that the policy reform process was still 'fresh'. This made it easier to investigate how the water sector reform was developed, who played the most important roles and which resources and discourses were applied. A timely policy evaluation profited from the availability of rich data in the form of informants who remembered the details and the available grey literature. Part of the reform could be studied while it was unfolding, using participatory observation.

This notwithstanding, the short time frame for evaluation (approx. 3 years after the reform) also has methodological disadvantages. Two important acts supporting the reform – WSIA and SPANA – had just been accepted and were still being implemented, while other subsidiary legislation, such as licensing regulations and permit rules had not been finalized. Not all the important institutions had been established. While the regulator and financier came into existence in 2006 and 2007 (respectively), they had yet to become fully operational. Several water corporations (such as SAMB and SAINS) had just begun to operate, while SADA did not yet exist. In addition, most water utilities were still struggling to establish proper information management systems for recording, retrieving, auditing, quality control and utilizing. It can be concluded that the short evaluation period, directly after the reform, made it difficult to draw evidence-based conclusions on the outcomes of the reform, although this was less of an issue when looking at the outputs.

Then, what would be the most appropriate time to conduct an evaluative research of a reform such as this? Should we wait until all the laws and institutions are stabilized and well-functioning and all the important information is available? The answer depends on the kind of questions to be answered. Arguably it is preferable not to rely on a one-off evaluation of such a major reform, but to evaluate it at two or three different moments in time. This evaluative study has managed to draw conclusions on, and explain how, the reform process took place and what outputs were formulated to meet the objectives. The reform measures and institutions will probably stabilize around 2012, when many other things will have fallen in place – the regulator will be properly equipped and have access to information, information management will be established, standardized performance indicators will be enforced on water utilities, and the remaining two state departments (JBA Labuan and LAP) will see their water suppliers corporatized, the PAAB's financial mechanism will have stabilized, the laws will have been fully implemented, etc. At this juncture I propose that a second evaluation covering the period 2007-2012 – when the reform has been fully functioning for five years – would be able to provide evidence-based conclusions on the outcomes of the water sector reform.

### 7.5.2 Data availability

This research clearly showed how the lack of available information hindered the outcome evaluations for several indicators (i.e. collection efficiency and water quality, see Chapters 5 and 6). In this research several problems with the availability of information were experienced as a result of data scarcity and data disclosure. The former describes the situation when data is really unavailable or partly available but inconsistent and suffering validity problems. The later occurred when respondents provided incomplete answers, refused to provide or disclose requested data or did not return questionnaires – without which the response rates of questionnaires (of 60%) would have been higher. This occurred even though the investigator is well-networked in the Malaysian water sector. Despite numerous follow-ups (through email and phone-calls), it was difficult to ascertain information on revenues, collection efficiency and unit production costs as some companies considered this data to be private and confidential. These problems would have been more serious if the researcher had been less well-connected. It is likely that those investigating the water sector in (other) developing countries (and more so in undeveloped countries) will experience similar methodological constraints bought about by a combination of economic constraints and political cultures. Political cultures of openness and disclosure might not help in producing more data, but are important in making existing data fully available. In Malaysia there is still much room for improvement in this respect. The Malaysian Centre for Independent Journalism (2007) reported that (environmental) information on water privatization, river pollution and drinking water quality are not (fully) available to the public. To some extent, the government treated this information as highly secret and used laws (i.e. Official Secrets Act 1972) to prohibit public access. Malaysia still faces the challenge of moving towards further information openness, disclosure and transparency.

### 7.5.3 The use of case studies

This research utilized two in-depth case studies to analyze the impacts of the reform on private and public water utilities ((PBAPP and SADA respectively). These case studies served two goals: (1) to better understand the causality between the reform process and outcomes; and (2) to better understand the relevance of public and private water utilities in explaining outcomes of the reform. These case studies produced interesting insights. They supported the general finding of this research about the weak correlation between the reform and the performance of water utilities. These case studies re-affirmed that the performance of water utilities was caused rather by external factors that existed before the reform was introduced.

Even though these case studies produced several useful insights, their application in this research had several limitations. First, the case studies were only used to examine the performance of water operators in terms of operational efficiency and environmental effectiveness. They did not attempt to explore the institutional reform process or the effectiveness of the reform's outputs. On reflection it may have been better to expand these case studies by including these areas, and applying the four dimensions of the PAA approach at different levels: the federal and state government levels as well as the private and public water utilities levels.

Second, the two selected case studies cannot be considered representative of each sector. For instance, the PBAPP is involved in treatment and distribution and is not representative of operators just involved in treatment or distribution. Meanwhile SADA was chosen to represent the corporatized public operator, but was not representative of public water departments. As explained, time and resource constraints limited the research to just two case studies, which did nonetheless make an important contribution to understanding and assessing the reform. Further research, using more case studies, might illustrate the repercussions of the reform on different categories of water operators.

## 7.6 Suggestions for future research

While this study has found a number of answers to the research questions, several new questions and challenges have emerged. Three promising areas for further research are outlined here.

First, as mentioned above the two case studies investigated do not cover the entire spectrum of public and private water utilities. A wider selection of case studies from both categories could be considered in future research on this topic, using both a PEA and a PAA approach. The second suggestion is to follow up this research after more time has lapsed. The 'interim' or 'work-in-progress' analysis (of 3 years) of this study proved to be too short to evaluate the societal impacts of the reform. As suggested in Section 7.5.1, a second evaluation research would be best undertaken when more (reliable and quality) information is available, the institutions are fully in place and performing well and the major regulations have been fully implemented.

The third and final suggestion is to assess the regulatory costs to the water utilities resulting from the reform. In addition to commercial risks (i.e. servicing bank loans), there are always costs implications for firms obliged to comply with new regulations (Parker, 2003). Usually these costs are factored into water prices and passed onto consumers. In countries where the water sector is well developed (e.g. the UK), regulatory costs are seen as the biggest business risk faced by water utilities and as such can have significant financial impacts to water utilities and end users. Such research could be carried out in the form of a comparative study (aided by case studies) to specifically examine the effects of regulatory costs on water utilities – between the private and public sectors – and water users – between domestic and industrial users. For water utilities, the impacts of the regulatory costs are related to (1) the proportion of regulatory in terms of their overall cost structure; and (2) how these regulatory costs can be managed. From the water users' perspective, an analysis of regulatory costs could help identify (1) the influence that they have on water tariffs (for both domestic and industrial customers) and (2) how water utilities implement of regulatory costs. Conducting such an evaluation would require developing a new set of parameters or indicators.

## 7.7 Policy recommendations

The reform was implemented as a public policy intervention by the federal government that aimed to improve the country's water supply sector. Since this study examined the on-going effects of the reform (*ex nunc* evaluation) it is possible to suggest policy directions that could strengthen the reform. In this regard, three major policy recommendations are worth considering.

First, there should be concrete and firm commitments from the government to ensure and facilitate public participation in decision making processes. This research has shown that the government recognizes the importance of civil society organizations (CSOs) and encouraged them to become involved in the reform process. Nevertheless, their involvement in the sector is still quite limited and they have not (yet) been able to balance the dominant role of state actors, nor the for-profit private sector. As we saw, CSOs made several proposals to amend the reform, but these were not accepted by the government. In the spirit of promoting good governance, transparency and democracy in the water sector (and more generally) the government should allow CSOs to play a significant role in the water sector and give them access to the information they need to do so. This will require amendments to the WSIA. Section 70 needs to be amended to protect the public platform, the Water Forum, against interference from market actors (i.e. water operators) and ensure that it remains a truly public platform. An amendment to Section 29 is also crucial to allow civil society to have better access to information held by the regulator and the water utilities. All the information relating to the functions of the NWSC and water utilities – annual reports, minutes of meetings, information on water quality, environmental performance, expenditures, etc. – should be made public (through new media such as the internet). This will help to hold both sets of institutions accountable to those who pay their salary, the public at large.

Another recommendation to emerge from this research is the call made by both water utilities and civil society for the government to streamline water management under a single body. At present, water management is un-coordinated – with several actors (at federal and state levels) having jurisdiction over this matter – creating a bureaucratic bottleneck to sustainable water management. This call is in line with the observation of Md. Khalid and Ab. Rahman (2010) that federal and state agencies have competing interests over water allocation for agriculture, fishery, forestry and irrigation. As this research indicated, consolidating water management under one 'water ministry' will not only bring all water-related agencies together, but also help harmonize policy frameworks, regulations and laws relating to water (currently enforced by different agencies). This is in line with the Integrated Water Resources Management approach and needs to include the sewerage sector (Chan, 2007; MEWC, 2008).

The main obstacles faced in conducting this research were the difficulty in getting required information from respondents and the reliability of the information available. This problem was partly caused by the absence of a clear policy relating to information management – recording, retrieving, validating and quality control in the water sector. The private water operators demonstrated some capability in managing information, but the public water utilities were having difficulties in installing such systems, mainly because of financial constraints. Moreover, no single body has been entrusted to coordinate this task. Hence, it is imperative that the government set up a national data bank or information depository for the water sector. The presence of such body would benefit the water sector in many ways: (1) it would enhance the effectiveness of the regulator, as regulation oversight works on the basis of information; (2) it would encourage the implementation of a benchmarking culture in the sector, since benchmarking cannot be implemented without (reliable) information; (3) it would give legitimacy to young and still fragile regulatory agencies, such as the NWSC (see Franceys & Gerlach, 2011); and (4) it would facilitate good governance and more democratic practices in the water sector. This task should be assigned to the NWSC, the sole regulatory body for the sector, and the institute best placed to carry out this task successfully. In taking on this task the NWSC should ensure that all stakeholders in the water sector are given equal access to information within the data bank and impose a compulsory requirement upon water utilities to provide information to this depository.

# References

Abbott, M. & Cohen, B. (2009). Productivity and efficiency in the water industry. *Utilities Policy*, 17(3-4): 233-244.

Abbott, M., Wang, W.C. & Cohen, B. (2011). The long-term reform of the water and wastewater industry: the case of Melbourne in Australia. *Utilities Policy*, 19(2): 115-122.

Agvin-Birikorang, S, Oladeji, O.O., O'Connor, G.A., Obreza, T.A. & Capece, J.C. (2009). Efficacy of drinking-water treatment residual in controlling off-side phosphorus losses: A field study in Florida. *Journal Environmental Quality*, 38(3): 1076-1085.

Ahlers, R. (2010). Fixing and nixing: the politics of water privatization. *Review of Radical Political Economics,* 42(2): 213-230.

Alegre, H., Baptista, J.M., Cabrere Jr. E., Cubillo, F., Duarte, P., Hirner, W., Merkel, W. & Parena, R. (2006). *Performance indicators for water supply services (2ⁿᵈ ed.)*. London, UK: IWA Publishing.

Aliran Ehsan Resources Berhad (2008). *Annual Report 2008*. Kuala Lumpur, Malaysia: Aliran Resources Berhad.

American Water Works Association (AWWA) (2004). *Selection and definition of performance indicators for water and wastewater utilities*. Denver, USA: AWWA Research Foundation and American Water Works Association.

American Water Works Association Research Foundation (AWWARF) (2007). *Advancing the science of water: AWWARF and water treatment residuals*. Available at: http://www.waterrf.orf/Research/ResearchTopics/StateofTheScienceReports/ResearchonResidualsfromWaterTreatment.

Aminudin, B. (2009). *Study on characteristic, treatment and disposal of drinking water treatment plant residue*. Paper presented at the Water Malaysia 2009, Kuala Lumpur, 19-21 May.

Ancarani, A. & Capaldo, G. (2001). Management of standardized public services: a comprehensive approach to quality assessment. *Managing Service Quality*, 11(5): 331-341.

Angkasa Consulting Services (2011). *Water supply*. Available at: http://www.acssb.com.my/acssb/watersupply1.htm.

Anonymous (2009). *Profiles of selected water operators*. Paper presented at the WaterLinks Forum, Bangkok, Thailand, 28-30 September.

Anwandter, L. & Ozuna, T.J. (2002). Can public sector reforms improve the efficiency of public water utilities? *Environment and Development Economics*, 7(4): 687-700.

Araral, E. (2009). The failure of water utilities privatization: synthesis of evidence, analysis and implications. *Policy and Society*, 27(3): 221-228.

Araral, E. (2010). Reform of water institutions: review of evidences and international experiences. *Water Policy*, 1(Suppl. 1): 8-22.

Arts, B. & Goverde, H. (2006). The governance capacity of (new) policy arrangements: a reflexive approach. In: B. Arts and P. Leroy (eds.), *Institutional dynamics in environmental governance*. Dordrecht, the Netherlands: Springer, pp. 67-72.

Arts, B. & Tatenhove, J.V. (2004). Policy and power: a conceptual framework between the 'old' and 'new' policy idioms. *Policy Science*, 37(3-4): 339-356.

Asian Development Bank (2000). *Developing best practices for promoting private sector investment in infrastructure*. Manila, Philippines: Asian Development Bank.

Asian Development Bank (2007). *Asian water development outlook 2007: achieving water security for Asia*. Manila, Philippines: Asian Development Bank.

Asian Development Bank (2008). *Good practices in urban water management: decoding good practices for a successful future*. Manila, Philippines: Asian Development Bank and Lee Kuan Yew School of Public Policy, National University of Singapore.

Asian Development Bank (2010). *Every drop counts: learning from good practices in eight Asian cities*. Manila, Philippines: Asian Development Bank and Institute of Water Policy, Lee Kuan Yew School of Public Policy, National University of Singapore.

Asian Development Bank, United Nations Development Programme, United Nations Economic and Social Commission for Asia and The Pacific and World Health Organisation (2006). *Asia water watch 2015: are countries in Asia on track to meet Target 10 of the Millennium Ddevelopment Goals?* (Publication Stock No. 020206). Manila, Philippines: Asian Development Bank.

Association of Dutch Water Companies (VEWIN) (2006). *Reflections of performance 2006: benchmarking in the Dutch drinking water industry*. Rijswijk, the Netherlands. Available at: www.vewin.nl.

Association of Dutch Water Companies (VEWIN) (2010). *Reflections of performance 2009: benchmarking in the Dutch drinking water industry*. Rijswijk, the Netherlands. Available at: www.vewin.nl.

Association of Water and Energy Research Malaysia (AWER) (2011). *National water service industry: Kelantan state case study*. Kuala Lumpur, Malaysia: AWER.

Barry, J. (2005). Ecological modernization. In: J.S. Dryzek and D. Schlosberg (eds.), *Debating the earth: the environmental politics readers (2^{nd} ed.)*. New York, USA: Oxford University Press, pp. 303-321.

Bartle, I. & Vass, P. (2007). Independent economic regulation: a reassessment of its role in sustainable development. *Utility Policy*, 15(4): 225-280.

Basta, N.T. & Dayton, E.A. (2001). Characterization of drinking water treatment residuals for use as a soil substitute. *Water Environment Research*, 73(1): 52-57.

Bennear, L.S. & Olmstead, S.M. (2008). The impacts of the 'right to know': information disclosure and the violation of drinking water standards. *Journal of Environmental Economics and Management*, 56: 117-130.

Berg, S.V. & Mugisha, S. (2010). Pro-poor water service strategies in developing countries: promoting justice in Uganda's urban project. *Water Policy*, 12(4): 589-601.

Bhuiyan, S.H. & Amagoh, F. (2011). Public sector reform in Kazakhstan: issues and perspective. *International Journal of Public Sector Management*, 24(3): 227-249.

Biswas, A. K. & Tortajada, C. (2010). Future water governance: problems and perspectives. *International Journal of Water Resources Management*, 26(2): 129-139.

Boag, G. & McDonald, D.A. (2010). A critical review of public-public partnerships in water services. *Water Alternatives*, 3(1): 1-25.

Boot, S. (2007). *Economic policy instruments and evaluation methods in Dutch water management* (PhD Dissertation). Rotterdam, the Netherlands: Erasmus University Rotterdam.

Brinton, S. R., O'Connor, G.A. & Oladeji, O.O. (2008). Surface applied water treatment residuals affect bioavailable phosphorus losses in Florida sands. *Journal of Environmental Management*, 88(4): 593-1600.

Brown, K., Ryan, N. & Parker, R. (2000). New modes of service delivery in the public sector: commercializing government: commercializing government services. *The International Journal of Public Sector Management*, 13(3): 206-221.

Budds, J. & McGranahan, G. (2003). Are the debates on water privatization missing the point? experiences from Africa, Asia and Latin America. *Environment and Urbanization*, 15(2): 87-105.

Buizer, M. (2008). *Worlds apart: interactions between local initiatives and established policy* (PhD Dissertation). Wageningen, the Netherlands: Wageningen University.

Casarin, A.A., Delfino, J.A. & Delfino, M.E. (2007). Failures in water reform: lesson from the Buenos Aires's concession. *Utility Policy*, 15(40): 234-247.

Cawangan Bekalan Air (1996). *Laporan tahunan Jabatan Kerja Raya*. Kuala Lumpur, Malaysia: Cawangan Bekalan Air.

Chan, N.W. (2007). Partnerships in integrated water resources management (IWRM) – a case study of Water Watch Penang. In: Chan, N.W. and Bouguerra, L. (eds.), *World citizens' assembly on water: towards global sustainability*. Penang, Malaysia: Water Watch Penang, pp. 47-56.

Chan, N.W. (2009). Issues and challenges in water governance in Malaysia. *Iranian Journal of Environmental Health Science and Engineering*, 6(3): 143-152.

Coalition Agaist Water Privatization (CAWP) (2006). *Memorandum to HE Minister of Energy, Water and Communications on proposed amendments on the Water Services Industry Bill (2005) and Suruhanjaya Perkhidmatan Air Negara (2005)*. Kuala Lumpur, Malaysia: CAWP.

Consumers'Assocaition of Penang (CAP) & Sahabat Alam Malaysia (SAM) (2006). *Comments on the Water Services Industry Bill*. Kuala Lumpur, Malaysia: CAP and SAM.

Cook, P. & Uchida, Y. (2008). The performance of privatized enterprises in developing countries. *Journal of Development Studies*, 44(9): 1342-1353.

Crabbe, A. & Leroy, P. (2008). *The handbook of environmental policy evaluation*. London: Earthscan.

Crase, L. & Gandhi V.P. (eds.). (2009). Reforming *institutions in water resources management: policy and performance for sustainable development*. London, UK: Earthscan.

Damazo, J. (2006). Philippines: in need of reforms and investments. *Asian Water*, 22(8): 12-15.

Denzin, N.K. (1989). *The research act: a theoretical introduction to sociological methods*. New York: McGraw-Hill.

Department of Environment (1974). *Environment Quality Act 1974*. Kuala Lumpur, Malaysia: Department of Environment.

Department of Environment (1979). *Environmental Quality (Sewage and Industrial Effluents) Regulations 1979*. Kuala Lumpur, Malaysia: Department of Environment, Malaysia.

Department Of Environment (2005). *Environmental Quality (Scheduled Wastes) Regulations 2005*. Putrajaya, Malaysia: Department of Environment, Malaysia.

Department of Environment (2009). *Malaysia Environmental Quality Report 2009*. Putrajaya, Malaysia: Department of Environment.

Dore, Mohamed H.I., Kushner, J. & Zumer, K. (2004). Privatization of water in the UK and France – what can we learn? *Utilities Policy*, 12(1): 41-50.

Dorsch, J.J. & Yasin, M.M. (1998). A framework for benchmarking in the public sector: literature review and directions for future research. *International Journal of Public Sector Management*, 11(2/3): 91-115.

Dunn, W.N. (2004). *Public policy analysis: an introduction*. New Jersey, USA: Pearson Prentice Hall.

Economic Planning Unit (2000). *National Water Resources Study 2000-2050*. Putrajaya, Malaysia: Economic Planning Unit.

Economic Planning Unit (2004). *Handbook: economic instruments for environmental management Malaysia*. Putrajaya, Malaysia: Economic Planning Unit.

Economic Planning Unit (2005). *Malaysia achieving the Millennium Development Goals: successes and challenges*. Putrajaya, Malaysia: Economic Planning Unit.

Economic Planning Unit (2006). *Ninth Malaysia Plan 2006-2010*. Putrajaya, Malaysia: Economic Planning Unit.

Economic Planning Unit (2008). *Mid-term review of Ninth Malaysia Plan 2006-2010*. Putrajaya, Malaysia: Economic Planning Unit.

Economic Planning Unit (2010). *Tenth Malaysia Plan 2011-2015*. Putrajaya, Malaysia: Economic Planning Unit.

Ehrhardt, D. & Janson, N. (2010). Can regulation improve the performance of government-controlled water utilities? *Water Policy*, 12(Suppl. 1): 23-40.

Estache, A. & Iimi, A. (2011). (Un)Bundling infrastructure procurement: evidence from water supply and sewage projects. *Utilities Policy*, 19(2): 104-114.

Farley, M., Wyeth, G., Md. Ghazali, Z. & Singh, S. (2008). *The manager's non-revenue water handbook: a guide to understanding water losses*. Kuala Lumpur, Malaysia: Ranhill Utilities Berhad and United States Agency for International Development (USAID).

Filbeck, G. & Gorman, R.F. (2004). The relationship between the environmental and financial performance of public utilities. *Environmental and Resources Economics*, 29: 137-157.

Fischer, F. (2006). Participatory governance as deliberative empowerment: the cultural politics of discursive space. *The American Review of Public Administration*, 36: 19-40.

Foster, V. (2005). *Ten years of water service reform in Latin America: toward an Anglo-French model* (Water Supply and Sanitation Sector Board Discussion Paper Series Paper No. 3). Washington, DC, USA: The World Bank Group. Available at: http://siteresources.worldbank.org/INTWSS/Resources/WSSServiceReform.pdf.

Franceys, R. & Weitz, A. (2003). Public-private community partnership in infrastructure for the poor. *Journal of International Development,* 15(8): 1083-1098.

Franceys, R.W.A. & Gerlach, E. (2011). Consumer involvement in water services regulations. *Utilities Policy*, 19(2): 61-70.

Fuest, V. & Haffner, S.A. (2007). PPP-policies, practices and problems in Ghana's urban water supply. *Water Policy*, 9(2): 169-192.

Gerlach, E. & Franceys, R. (2010). Regulating water services for all in developing economies. *World Development*, 38(9): 1229-1240.

Giddens, A. (1984). *The constitution of society*. Cambridge, UK: Polity Press.

Gysen, J., Bruyninckx, H. & Bachus, K. (2006). The modus narrandi: a methodology for evaluating effects of environmental policy. *Evaluation,* 12(1): 95-118.

Hadipuro, W. (2010). Indonesia's water supply regulatory framework: between commercialization and public service? *Water Alternatives*, 3(3): 475-491.

Hajer, M.A. (1995). *The politics of environmental discourse: ecological modernization and the policy process*. Oxford, UK: Oxford University Press.

Hall, D. & Lobina, E. (2006). *Water as a public service*. A Report Commissioned by Public Service International Research Unit, University of Greenwich. Available at: http://www.world-psi.org.

Hall, D., Corral, V., Lobina, E. & De la Motte, R. (2004). *Water privatization and restructuring in Asia-Pacific*. London, UK: Public Service International Research Unit (PSIRU), University of Greenwich.

Hall, D., Lobina, E., Corral, V., Hoedeman, O., Terhorst, P., Pigeon, M. & Kishimoto, S. (2009). *Public-public partnerships (PUPs) in water*. London, UK: Public Service International Research Unit (PSIRU), University of Greenwich.

Hardoy, A. & Schusterman, R. (2000). New models for the privatization of water and sanitation for the urban poor. *Environment and Urbanization*, 12(63): 63-75.

Harris, S. (2003). *Public private partnerships: delivering better infrastructure* (Working Paper). Washington, DC, USA: Inter-American Development Bank.

Hassan, N.I. (2006). *Characterisation of water treatment sludge* (Unpublished B. Eng. (Hons) thesis). Malaysia: Universiti Teknologi Mara.

Hayward, K. (2007). The secrets of a successful public water utility. *Water Utility Management International*, 2(1): 5-17.

Hemming, P.J. (2008). Mixing qualitative research methods in children's geographies. *Area*, 40(2): 152-162.

Holland, A. S. (2005). *The water business: corporatization versus people*. London, UK: Zed Books.

Hsieh, H.-N. & Raghu, D. (2008). Characterization of water treatment residuals and their beneficial uses. *Proceedings of the GeoCongress 2007: geotechnics of waste management and remediation, American Society of Civil Engineers, USA*.

Hua, T.Y. (2009). Water sector reforms in Malaysia. *Water Utility Management International*, 4(1): 16-17.

Huang, C., Pan, J.R., Sun, K.D. & Liaw, C.T. (2001). Reuse of water treatment plant sludge and dam sediment in brick making. *Water Science and Technology*, 44(10): 273-277.

Hukka, J.J. & Katko, T.P. (2003). Refuting the paradigm of water services privatization. *Natural Resources Forum*, 27: 142-135.

Iacob, C. & Farcas, A. (2010). *Water treatment plant residuals management*. Available at: http://www.yrc.utcb.ro/p/YRC_2010_GH_Cristina_Iacob.pdf.

Ippolito, J.A., Barbarick, K.A. & Elliot, H.A. (2011). Drinking water treatment residuals: a review of recent uses. *Journal of Environmental Quality*, 40(1): 1-12.

Jabatan Audit Negara (2008). *Laporan Ketua Audit Negara tahun 2008: pengurusan kualiti air minum seluruh negara*. Putrajaya, Malaysia: Jabatan Audit Negara.

Jabatan Bekalan Air (2012). *A glimpse at water supply in Malaysia: past and present*. Putrajaya, Malaysia: Jabatan Bekalan Air. Available at: http://www.jba.gov.my/index.php/bm/semenanjung-malaysia.

Jaseni, M. (2009). *Holistic NRW management*. Paper presented at the Water Malaysia 2009, Kuala Lumpur, 19-21 May.

Jerome, A. (2006). Privatization and regulation in South Africa: an evaluation. In: Amann, E. (ed.), *Regulating development*. Cheltenham, UK: Edward Elgar, pp. 179-197.

Kamat, S. (2004). The privatization of public interest: theorizing NGO discourse in a neoliberal era. *Review of International Political Economy*, 11(1): 155-176.

Kessides, I.N. (2004). *Reforming infrastructure: privatization, regulation and competition*. Washington, DC, USA: World Bank and Oxford University Press.

Khoo, T.C. (2007). *The Singapore water history*. Paper presented at the 6th Ministers' Forum on infrastructure development in Asia and Pacific region, Beijing, PRC, 28-29 August.

Kibassa, D. (2011). The impact of cost recovery and sharing system on water policy implementation and human right to water: a case of Ileje, Tanzania. *Water Science and Technology*, 63(11): 2520-2526.

Kikeri, S. & Nellis, J. (2004). An assessment of privatization. *The World Bank Research Observer*, 19(1): 87-118.

Kirkpatrick, C., Parker, D. & Zhang, Y.F. (2006). A comparative analysis of the performance of public and private water utilities in Africa. In: Amman. E. (ed.), *Regulating development: evidence from Africa and Latin America*. Cheltenham, UK: Edward Elger. pp. 198-221.

Klostermann, J.E.M. & Cramer, J. (2007). Social Construction of sustainability in water companies in the Dutch coastal zone. *Journal of Cleaner Production*, 15(16): 1573-1584.

Kuks, S.M.M. (2006). The privatization debate on water services in the Netherlands: public performance of the water sector and the implications of market forces. *Water Policy*, 8(2): 147-169.

Kumar, R. (2005). *Research methodology: a step-by-step guide for beginners* (2nd ed.). London, UK: Sage.

Kvale, S. (1996). *Interviews*. London, UK: Sage.

Lambert, A.O., Brown, T.G., Takizawa, M. & Weimer, D. (1999). A review of performance indicators for real losses from water supply systems. *Journal of Water Supply: Research and Technology – AQUA*, 48(6): 227-237.

Lee, E. (2010). Information disclosure and environmental regulation: green lights and grey areas. *Regulation and Governance*, 4: 303-328.

Lembaga Air Perak (LAP) (2011). *Pemberian insentif kepada pemberi maklumat kes kecurian air*. Available at: http://www.lap.com.my/web/.

Liefferink, D. (2006). The dynamics of policy arrangements: turning round the tetrahedron. In: B. Arts and P. Leroy (eds.), *Institutional dynamics in environmental governance*. Dordrecht, the Netherlands: Springer, pp. 45-68.

Linjun, Z. (2007). *Regulation on entry and after entry*. Paper presented at the Global Water 2007 Conference, Barcelona, Spain, 2-3 April.

Lobina, E. & Hall, D. (2007). Experience with private sector participation in Grenoble, France and lesson on strengthening public water operations. *Utilities Policy*, 15(2): 93-109.

Makris, K.C. & O'Connor, G.A. (2007). Beneficial utilization of drinking-water treatment residuals as contaminant-mitigating agents. In: Sarkar, D., Datta, R. and Hannigan, R. (eds.), *Current perspectives in environmental geochemistry*. Denver, USA: Geological Society of America Press, pp. 609-635.

Malaysia Nature Society (MNS) (2009). *The proposal on the need for a water demand management plan in Selangor submitted to the state EXCO of Selangor for Tourism, Consumer Affairs and Environment*. Kuala Lumpur, Malaysia: MNS.

Malaysia Water Association (MWA) (2003). Malaysia W*ater Industry Guide 2003*. Kuala Lumpur, Malaysia: Malaysia Water Association.

Malaysia Water Association (MWA) (2005). Malaysia W*ater Industry Guide 2005*. Kuala Lumpur, Malaysia: Malaysia Water Association.

Malaysia Water Association (MWA) (2006). Malaysia W*ater Industry Guide 2006*. Kuala Lumpur, Malaysia: Malaysia Water Association.

Malaysia Water Association (MWA) (2007). Malaysia W*ater Industry Guide 2007*. Kuala Lumpur, Malaysia: Malaysia Water Association.

Malaysia Water Association (MWA) (2008a). *Malaysia Water Industry Guide 2008*. Kuala Lumpur, Malaysia: Malaysia Water Association.

Malaysia Water Association (MWA) (2008b). History and perspective of Malaysian water industry development in Penang (water section). *Water Malaysia*, 16: 6-8.

Malaysia Water Association (MWA) (2008c). *Study on characteristic, treatment and disposal of potable water treatment plant residue: stage 1 literature review*. Kuala Lumpur, Malaysia: Malaysia Water Association.

Malaysia Water Association (MWA) (2009). *Malaysia Water Industry Guide 2009*. Kuala Lumpur, Malaysia: Malaysia Water Association.

Malaysia Water Association (MWA) (2010). *Malaysia Water Industry Guide 2010*. Kuala Lumpur, Malaysia: Malaysia Water Association.

Malaysia Water Association (MWA) (2011). *Malaysia Water Industry Guide 2011*. Kuala Lumpur, Malaysia: Malaysia Water Association.

Marques, R.C. (2006). A yardstick competition model for Portuguese water and sewerage services regulation. *Utilities Policy*, 14(3): 175-184.

Marsden Jacob Associates (2005). *Third party access in water and sewerage infrastructure: implications for Australia* (Research Paper). Australia: Department of Agriculture, Fisheries and Forestry, Government of Australia.

Massarutto, A. (2007). *Liberation and private sector involvement in the water industry: a review of the economic literature*. (Working Paper Series). Milan, Italy: Centre for Research on Energy and Environmental Economics and Policy, Bocconi University.

McDonald, D.A. (2012). Remunicipalisation works. In: Pigeon, M., McDonald, D.A., Hoedeman, O. and Kishimoto, S. (eds.), *Municipalisation: putting water back into public hands.* Amsterdam, the Netherlands: Transnational Institute, pp. 8-23.

Mckenzie, R. & Lambert, A. (2002). *ECONOLEAK: economic model for leakage management for water suppliers in South Africa: user guide* (WRC Report TT 169/02). Republic of South Africa: South African Water Research Commission.

McKenzie, R. & Seago, C. (2005). Assessment of real losses in potable water distribution systems; some recent developments. *Water Science and Technology: Water Supply,* 5(1): 33-40.

Md Khalid, R. & Ab Rahman, S. (2010). *Legal analysis of sustainable development and water management in Malaysia.* Paper presented at the 16th International Sustainable Development Research Conference, Hong Kong SAR, 30 May – 1 June.

Mei, K.Y. (2009). Court: make water deal public. *The Star* June 29, 2009. Available at: http://www.thestar.com.my.

Meijer, C.P., Verloop, N. & Beijaard, D. (2002). Multi-method triangulation in qualitative study on teacher's practical knowledge: an attempt to increase internal validity. *Quality and Quantity,* 36(2): 145-167.

Ministry of Energy, Water and Communications (MEWC) (2004). *Study on the development of the water supply industry and the establishment of a national water supply commission.* Kuala Lumpur, Malaysia: The Ministry of Energy, Water and Communications.

Ministry of Energy, Water and Communications (MEWC) (2006a). The *Water Services Industry Act (Act 655).* Putrajaya, Malaysia: Ministry of Energy, Water and Communications.

Ministry of Energy, Water and Communications (MEWC) (2006b). *Suruhanjaya Perkhidmatan Air Negara Act (Act 654).* Putrajaya, Malaysia: Ministry of Energy, Water and Communications.

Ministry of Energy, Water and Communications (MEWC) (2007). *Orders, regulations and rules under the Water Services Industry Act 2006.* Putrajaya: Ministry of Energy, Water and Communications, Malaysia.

Ministry of Energy, Water and Communications (MEWC) (2008). *The water tablet: Malaysian water reforms.* Putrajaya, Malaysia: The Ministry of Energy, Water and Communications.

Ministry of Energy, Water and Communications (MEWC) (2010). *Projek penyaluran air mentah Pahang-Selangor.* Available at: http://www.kettha.gov.my/content/pasukan-projek-penyaluran-air-mentah-pahang-selangor-ppamps.

Ministry of Health (2008). *The National Guideline for Drinking Water Standard 2001.* Putrajaya, Malaysia: Ministry of Health.

MMC-Sumitomo Corporation (2011). *National water services master plan.* A joint study between Sumitomo Corporation and MMC Corporation Berhad in collaboration with Tokyo Metropolitan Government.

Modell, S. (2005). Triangulation between case study and survey methods in management accounting research: an assessment of validity implication. *Management Accounting Research,* 16(2): 231-254.

Mol, A.P.J. & Spaargaren, G. (1993). Environment, modernity and the risk society: the apocalyptic horizon of environmental reform. *International Sociology,* 8(4): 431-459.

Mol, A.P.J. (1995). *The refinement of production: ecological modernization theory and the chemical industry.* Utrecht, the Netherlands: Van Arkel.

Mol, A.P.J. (2008). *Environmental reform in the information age: the contour of information governance.* Cambridge: Cambridge University Press.

Mol, A.P.J. (2009). Environmental governance through information: China and Vietnam, *Singapore Journal of Tropical Geography,* 30(1): 114-129.

Mustafa, D. & Reeder, P. (2009). 'People is all that is left to privatize': water supply privatization, globalization and social justice in Belize City, Belize, *International Journal of Urban and Regional Research*, 33(3): 789-808.

Nickson, A. (2000). *The role of the non-the state sector in urban water supply*. Birmingham, UK: International Development Department, The University of Birmingham.

Niehof, A. (1999). *Household, family and nutrition research: writing a proposal*. Wageningen H and C Series 1. Wageningen, the Netherlands: Wageningen University.

Nigam, A. & Rasheed, S. (1998). *Financing of fresh water for all: a right based approach* (Staff Working Papers – evaluation, policy and planning series). New York, USA: UNICEF.

Norton, S.D. & Blanco, L. (2009). Public-private partnerships: a comparative study of new public management and stakeholder participation in the UK and Spain. *International Journal of Public Policy*, 4(3/4): 214-23.

Novak, J.M. & Watts, D.W. (2005). An alum-based water treatment residual can reduce extractable phosphorus concentrations in three phosphorus-enriched coastal plain soils. *Journal Environmental Quality*, 34(5): 1820-1827.

Nyarko, K.B. (2007). *Drinking water sector in Ghana: drivers for performance*. (PhD Dissertation). Delft, the Netherlands: UNESCO-IHE Institute for Water Education.

Othman, F. (2007). *Combining concession contract with BOT: challenges for the water operator: Ranhill Utilities experience in Malaysia*. Paper presented at the Global Water 2007 Conference, Barcelona, Spain, 2-3 April.

Owen, G. (2006). Sustainable development duties: new roles of UK economic regulators. *Utility Policy*, 14(3): 135-218.

Page, B. & Bakker, K. (2005). Water governance and water users in a privatized water industry: participation in policy-making and in water services provision: a case study of England and Wales. *International Journal of Water*, 3(1): 38-60.

Parker, D. (2003). The dynamics of regulation: performance, risk and strategy in the privatized, regulated industries. In: Wubben, E.F.M. and Hulsink, W. (eds.), *On creating competition and strategic restructuring: regulatory reform in public utilities*. Massachusetts, USA: Edward Elgar, pp. 69-97.

Patton, M.Q. (1999). Enhancing the quality and credibility of qualitative analysis. *Health Services Research*, 34(5): 1189-1208.

Pearce-Oroz, G. (2006). The viability of decentralised water and sanitation provision in developing countries: the case of Honduras. *Water Policy*, 8(1): 31-50.

Pengurusan Aset Air Berhad (PAAB) (2007). *Executive briefing to the Secretary-General of the Ministry of Energy, Water and Communications*. Kuala Lumpur, Malaysia: Pengurusan Aset Air Berhad.

Pengurusan Aset Air Berhad (PAAB) (2009). *Report on literature study and laboratory testing on characteristics of water treatment plants residuals for DOE consideration to declassify residuals as non-scheduled waste under Environment Quality Act 1974*. Kuala Lumpur, Malaysia: Pengurusan Aset Air Berhad.

Pengurusan Aset Air Berhad (PAAB) (2010). *Annual Report 2010*. Kuala Lumpur, Malaysia: Pengurusan Aset Air Berhad.

Pengurusan Aset Air Berhad (PAAB) (2011). *Custodian of national water assets*. Available at: http://www.paab.my/index.php.

Perard, E. (2009). Water Supply: Public or private? an approach based on cost of funds, transaction costs, efficiency and political costs. *Policy and Society*, 27(3): 193-219.

Perbadanan Bekalan Air Holdings Berhad (PBAHB) (2008). *Annual Report 2008*. Penang, Malaysia: PBA Holdings Berhad.

Perbadanan Bekalan Air Holdings Berhad (PBAHB) (2009). *Annual Report 2009*. Penang, Malaysia: PBA Holdings Berhad.

Perbadanan Bekalan Air Holdings Berhad (PBAHB) (2010). *Annual Report 2010*. Penang, Malaysia: PBA Holdings Berhad.

Perbadanan Bekalan Air Pulau Pinang (PBAPP) (2010). *Key performance indicator 2010*. Penang, Malaysia: PBAPP.

Perbadanan Bekalan Air Pulau Pinang (PBAPP) (2011). *Key performance indicator 2011*. Available at: http://www.pba.com.my/statistics.aspx.

Pierre, J. & Peters, B. G. (2000). *Governance, politic and the state*. Hampshire, UK: Macmillan Distribution Ltd.

Pigeon, M. (2012). Soggy politics: making water 'public' in Malaysia. In: Pigeon, M., McDonald, D.A., Hoedeman, O. and Kishimoto, S. (eds.), *Municipalisation: putting water back into public hands*. Amsterdam, the Netherlands: Transnational Institute, pp. 90-104.

Polidado, C. & Hulme, D. (1999). Public management reform in developing countries. *Public Management Review*, 1(1): 121-132.

Pongsiri, N. (2002). Regulation and public-private partnerships. *The International Journal of Public Sector Management*, 15(6): 487-495.

Prasad, N. (2006). Privatization results: private sector participation in water services after 15 years. *Development Policy Review*, 24(6): 669-692.

Prasad, N. (2007a). Privatisation of water: a historical perspective. *Law, Environment and Development Journal*, 3(2): 217-233.

Prasad, N. (2007b). *Social policies and water sector reform* (Markets, Business and Regulation Programme Paper Number 3). Genera, Switzerland: UNRISD.

Public Utility Berhad (PUB) (2010). *Annual Report 2009/2010*. Singapore: PUB.

Puncak Niaga Holdings Berhad (PNHB) (2008). *Annual Report 2008*. Kuala Lumpur, Malaysia: Puncak Niaga Holdings Berhad.

Puncak Niaga Holdings Berhad (PNHB) (2009). *Annual Report 2009*. Kuala Lumpur, Malaysia: Puncak Niaga Holdings Berhad.

Puncak Niaga Holdings Berhad (PNHB) (2010). *Annual Report 2010*. Kuala Lumpur: Puncak Niaga Holdings Berhad.

Ranhill Utilities Berhad (2007). *Annual Report 2007*. Kuala Lumpur, Malaysia: Ranhill Utilities Berhad.

Ranhill Utilities Berhad (2008). *Sustainability Report 2008*. Kuala Lumpur, Malaysia: Ranhill Utilities Berhad.

Ranhill Utilities Berhad (2009). *Annual Report 2009*. Kuala Lumpur, Malaysia: Ranhill Utilities Berhad.

Ranhill Utilities Berhad (2011). *Annual Report 2011*. Kuala Lumpur, Malaysia: Ranhill Utilities Berhad.

Reddy, R.V. (1998). Institutional imperatives and co-production strategies for large irrigation systems in India. *Indian Journal of Agricultural Economics*, 53(3): 440-455.

Rhodes, R.A.W. (1986). *The national world of local government*. London, UK: Allen and Unwin.

Robbins, P.T. (2003). Transnational corporations and the discourse of water privatization. *Journal of International Development*, 15(8): 1073-1082.

Rogers, P., De Silva, R. & Bhatia, R. (2002). Water is an economic good: how to use prices to promote equity, efficiency and sustainability. *Water Policy*, 4: 1-17.

Rouse, M. (2007). *Institutional governance and regulation of water services: the essential elements*. London, UK: IWA Publishing.

Sahabat Alam Malaysia (2004). *Malaysian environment in crisis*. Penang, Malaysia: Sahabat Alam Malaysia.

Salcon Engineering (2008). *Annual Report 2008*. Kuala Lumpur, Malaysia: Salcon Engineering.

Schwartz, K. (2008). The new public management: the future for reforms in the Africa water supply and sanitation sector? *Utilities Policy*, 16(1): 49-58.

Siddiquee, N.A. (2010). Managing for results: lessons from public management reform in Malaysia. *International Journal of Public Service Management*, 23(1): 38-53.

Silvestre, H.C. (2012). Public-private partnership and corporate public sector organizations: alternative ways to increase social performance in the Portuguese water sector? *Utilities Policy*, 22: 41-49.

Smith, M., Khadilah, Y. & Ahmad, M.A. (2007). Environmental disclosure and performance reporting in Malaysia. *Asian Review of Accounting*, 15(2): 185-199.

Sohail, M. (2000). *PPP and the poor in water and sanitation: interim findings on PPP case study in Queenstown, South Africa*. Water, Engineering and Development Centre. Loughborough, UK: Loughborough University.

Spaargaren, G. & Mol, A.P.J. (1992). Sociology, environment and modernity: ecological modernization as a theory of change. *Society and Natural Resources*, 5(4): 323-344.

Spiller, P.T. & Savedoff, W. (1997). Commitment and governance in infrastructure sector. In Willig, Uribe and Basanes (eds.), *Can privatization deliver? infrastructure for Latin America*. Baltimore, MD, USA: Johns Hopkins University Press, pp. 133-150.

Stedman, L. (2009). Structuring tariffs: getting the most from water rates. *Water Utility Management International*, 2(1): 20-21.

Stephan, M. (2002). Environmental information disclosure programs: they work, but why? *Social Science Quarterly*, 83(1): 190-205.

Stern, J. & Holder, S. (1999). Regulatory governance: criteria for assessing the performance of regulatory system: an application to infrastructure industries in the developing countries of Asia. *Utility Policy*, 8(1): 1-73.

Stern, J. (2000). Electricity and telecommunication regulatory institutions in small and developing countries. *Utilities Policy*, 9(3): 131-157.

Suruhanjaya Perkhidmatan Air Negara (SPAN) (2009). *Annual Report 2009*. Cyberjaya, Malaysia: Suruhanjaya Perkhidmatan Air Negara.

Suruhanjaya Perkhidmatan Air Negara (SPAN) (2010a). *Annual Report 2010*. Cyberjaya, Malaysia: Suruhanjaya Perkhidmatan Air Negara.

Suruhanjaya Perkhidmatan Air Negara (SPAN) (2010b). *Water services industry performance report 2010*. Cyberjaya Malaysia: Suruhanjaya Perkhidmatan Air Negara.

Suruhanjaya Perkhidmatan Air Negara (SPAN) (2011a). Asset light model. *Buletin SPAN*, 1: 12.

Suruhanjaya Perkhidmatan Air Negara (SPAN) (2011b). *Briefing note to the SWAn*. Cyberjaya, Malaysia: Suruhanjaya Perkhidmatan Air Negara.

Syabas (2005). *Syabas's briefing note to the Ministry of Energy, Water and Communications*. Kuala Lumpur, Malaysia: Syabas.

Syarikat Air Darul Aman (SADA) (2010). *Key performance indicator*. Alor Setar: Syarikat Air Darulaman Berhad.

Syarikat Air Darul Aman (SADA) (2011a). *Polisi syarikat*. Available at: http://www.sada.com.my/?page_id=1259.

Syarikat Air Darul Aman (SADA) (2011b). *Dasar kualiti*. Available at: http://www.sada.com.my/?page_id=1259.

Syed Zin, S.F. (2007). *Reclamation of water treatment sludge as a ceramic product* (Unpublished B. Eng. (Hons) thesis). Malaysia: Universiti Teknologi Mara.

Tasman Asia Pacific (1997). *Third party access in the water industry: an assessment of the extent to which services provided by water facilities meet the criteria for declaration of access* (Report). National Competition Council of the Government of Australia and Tasman Asia Pacific.

Tay, J.H., Show, K.Y. & Hong, S.Y. (2001). Reuse of industrial sludge as industrial aggregates. *Water Science and Technology*, 44(10): 269-272.

The Association of Chartered Certified Accountants (2002). *The state of corporate environmental reporting in Malaysia*. London, UK: Certified Accountants Educational Trust.

The Solid Waste and Public Cleaning Corporation (2010). *Annual Report 2010*. Kuala Lumpur, Malaysia: The Solid Waste and Public Cleaning Corporation.

Thompson, P. (2002). *Corporate environmental reporting in Singapore and Malaysia: progress and prospects*. Research Paper Series, No. 11/2002. Nottingham, UK: Centre for Europe-Asia Business Research.

Trafford, S. & Proctor, T. (2006). Successful joint venture partnerships: public-private partnerships. *International Journal of Public Sector Management*, 19(2): 17-129.

United Nations Development Program (UNDP) (2006). Human development report 2006. New York, USA: UNDP. Available at: http://hdr.undp.org/en/media/HDR06-complete.pdf.

United States Agency for International Development (USAID) (2005). *Case studies of bankable water and sewerage utilities* (Volume 1: overview report). Available at: http://pdf.usaid.gov/pdf_docs/PNADE147.pdf.

Van Tatenhove, J.P.M. & Leroy, P. (2003). Environmental and participation in a context of political modernization. *Environmental Values*, 12(2): 155-174.

Van Tatenhove, J.P.M., Arts, B. & Leroy, P. (eds.). (2000). *Political modernization and the environment: the renewal of environmental policy arrangements*. Dordrecht, the Netherlands: Kluwer Academic Publishers.

Van Vliet, B. (2002). *Greening the grid: the ecological modernisation of network-bound systems* (PhD Dissertation). Wageningen, the Netherlands: Wageningen University.

Vaz, S.G., Martin, J., Wilkinson, D. & Newcombe, J. (2001). *Reporting on environmental measures: are we being effective?* (Environmental Issues Report No. 25). Copenhagen, Denmark: European Environment Agency.

Wahid, M.A., Kamaruddin, K., Adam, T.A., Syed Zin, S.F., Hassan, N.I., Wan Jusof, W.A., Mohd Tajuddin, R., Jaafar, J. & Baki, A. (2008). Reuse of water treatment residue/sludge as pottery product. *Water Malaysia*, 16: 20-23.

Walsh, M.E., Lake, C.B. & Gagnon, G.A. (2008). Strategic pathways for sustainable management of water treatment plant residuals. *Journal of Environmental Engineering and Science*, 7(1): 45-52.

Wan Jusoh, W.A. (2007). *To study physical properties of water treatment sludge as a ceramic product*. (Unpublished B. Eng. (Hons) thesis). Malaysia: Universiti Teknologi Mara.

Warta Darulaman (2011). *Rakan SADA bantu tingkatkan perkhidmatan air*. Available at: http://www.wartakedah. net/index.php?option=com_content&view=article&id=941%3Arakan-sada-bantu-tingkatkan-perkhidmatan-ai r&catid=36%3Aberita&Itemid=67&lang=ms.

Weber, R.P. (1990). *Basic content analysis: quantitative applications in the social science*. Newbury Park, CA, USA: Sage.

Weil, D., Fung, A., Graham, M. & Fagotto, E. (2006). The effectiveness of regulatory disclosure policies. *Journal of Policy Analysis and Management*, 25(1): 155-181.

Wiering, M.A. & Arts, B.J.M. (2006). Discursive shifts in Dutch river management: 'deep' institutional change or adaptation strategy? *Hydrobiologia*, 565(1): 327-338.

Wiering, M.A. & Crabbe, A. (2006). The institutional dynamics of water management in the low countries. In: Arts, B. and Leroy, P. (eds.), *Institutional dynamics in environmental governance*. Dordrecht, the Netherlands: Springer, pp. 93-114.

Winarni, M. (2009). Infrastructure Leakage Index (ILI) as water losses indicator. *Civil Engineering Dimension*, 11(2): 126-134.

Winchester, H.P.M. (2005). Qualitative research and its place in human geography. In: Hay, I. (ed.) (2nd ed.), *Qualitative research methods in human geography*. Melbourne, Australia: Oxford University Press, pp. 3-18.

World Bank (2008). Thailand infrastructure Annual Report 2008. Available at: http://siteresources.worldbank.org/ INTTHAIL&/Resources/333200-1177475763598/3714275-1234408023295/5826366-1234408105311/chapter5-water-sanitation-&-low-income-housing-sector.pdf.

World Bank (2012). *Organic water pollutant (BOD) emissions (kg per day)*. Available at: http://data.worldbank.org/ indicator/EE.BOD.TOTL.KG/countries.

World Health Organization & United Nations International Children's Emergency Fund (WHO/UNICEF) (2010). *Progress on sanitation and drinking-water: 2010 update*. Geneva, Switzerland: WHO and UNICEF.

Xia, J. (2010). New opportunities and challenges on integrated water supply and water demand managements. *Journal of Resources and Ecology*, 1(3): 193-201.

Yin, R.K. (1994). *Case study research: design and methods* (2nd ed.). Thousand Oaks, CA, USA: Sage.

Yin, R.K. (2003). *Case study research: design and methods* (3nd ed.). Thousand Oaks, CA, USA: Sage.

Yin, R.K. (2009). *case study research: design and methods* (4th ed.). Thousand Oaks, CA, USA: Sage.

Yusoff, H., Lehman, G. & Nasir, N.M. (2006). Environmental engagement through the lens of disclosure practices: a Malaysian story. *Asian Review of Accounting*, 14(2): 122-148.

Yusoff, H., Yatim, N. & Nasir, N. (2004). *Analysis on the development of environmental disclosure practices by selected Malaysian companies from 1999 to 2002*. Paper presented at the Fourth Asia Pacific Interdisciplinary Research in Accounting Conference, Singapore, 4-6 July.

Zaharaton Raja Zainal Abidin (2004). *Water resources management in Malaysia: the way forward*. Paper presented at the Asia Water 2004, Kuala Lumpur, 30 Mar – 2 April.

Zaini, U., Rahman, R.A. & Anuar, A.N. (2008). *Current trends in water quality and resources management*. Proceedings of the 1st technical meeting of Muslim water researchers cooperation (MUWARWC), Kuala Lumpur, Malaysia, 15-16 December.

Zaini, U. (2009). *Water sustainability and strategic initiatives: a proposed water industry roadmap 2010-2020*. Keynote address presented at the Malaysia Water Conference and Exhibition, Kuala Lumpur, Malaysia, 19-21 May.

Zetland, D. (2011). *The end of abundance: economic solutions to water scarcity*. Amsterdam, the Netherlands: Aguanomics Press.

Zhong, L. (2007). *Governing urban water flows in China* (PhD Dissertation). Wageningen, the Netherlands: Wageningen University.

# Appendices

## Appendix 1. In-depth interview questions

### A. *Water operators*

1.  As one of the main stakeholders in the sector, what do you envisage from the reform?
2.  To what extent did the reform address the interests of water operators?
3.  How can water operators benefitted from the reform?
4.  What were your top most concerns about the reform?
5.  Were they being addressed by the reform?
6.  Were you invited to participate in the reform?
7.  If no, why? If yes, what did you propose to the government?
8.  To what degree did you manage to convince the government to consider your suggestions?
9.  If no, what are the main obstacles?
10. Did you work alone or collaborate with other parties to convince the government?
11. If no, why? If yes, who were they? Were they effective? If no, why?
12. Can NWSC enforce WSIA 2006 effectively to ensure that water operators are able to discharge their duty in an efficient and effective manner?
13. If no, what prohibits NWSC from discharging its responsibility effectively? If yes, what have NWSC done with regard to water supply sector in Malaysia generally?
14. Do you happy/satisfy with the performance NWSC thus far?
15. If no, why? If yes, please state in what areas?
16. Please suggest how can NWSC or the government could facilitate efficient and effective water supply management in Malaysia?
17. Do you see that certain provisions in WSIA 2006 have to be amended to bring about greater efficiency and effectiveness to the sector?
18. If no, why? If yes, what amendments do you suggest? How can they help to improve the overall sector?
19. From the perspective of water operators, what are crucial policy interventions from government to ensure water supply sector in Malaysia is sustainable?
20. What other aspects or measures do you want to propose to be included in the reform in the future?
21. How can they be implemented?
22. To what extent do you consider environmental issues in the decision-making process in the company?
23. What is your opinion about incorporating 'environment tax/green tax' in the water tariff structure?
24. What is the state of your WTP? Do they have sludge treatment facilities? If no where do you dispose the sludge?
25. Do you think who should be responsible for sludge treatment? Water operators or government?

## B. *Consumer associations*

1.  How can the reform benefit consumers in general?
2.  What were your top most concerns about the water supply sector in Malaysia?
3.  Were you invited to participate during the reform peocess?
4.  If no, why? If yes, what did you propose?
5.  To what degree did you manage to convince the government?
6.  If no, what were the main obstacles?
7.  Did you work alone or collaborate with other parties?
8.  If no, why? If yes, who were they? Were they effective? If no, why?
9.  In your opinion, what is the function of consumers' involvement in the water sector?
10. From the perspective of consumer, what are crucial policy interventions from government to ensure water supply sector in Malaysia is sustainable?
11. Do you think regulator (NWSC) be able to enforce WSIA 2006 effectively, including giving due consideration to consumers interests?
12. If no, what prohibits NWSC from discharging its responsibility effectively? If yes, what have NWSC done with regard to protecting consumer interests?
13. Please suggest how can NWSC facilitate or encourage better consumer participations in the decision making of the sector?
14. How can Water Forum be used to promote the consumer interests in the sector?
15. Please suggest how can consumers association (like FOMCA/CAP/PCPA) play important roles in the water sector?

## C. *Environmental organizations*

1.  How can the reform benefit the environment in general?
2.  What were your top most concerns about the water supply sector in Malaysia from the perspective of environment?
3.  Were you invited to give feedback during the reform process?
4.  If no, why? If yes, what did you propose?
5.  To what degree did you manage to convince the government?
6.  If no, what were the main obstacles?
7.  Did you work alone or collaborate with other parties to convince the government?
8.  If no, why? If yes, who were they? Were they effective? If no, why?
9.  In your opinion, how's important environmentalists' involvement in the water sector? If yes, why is it important?
10. From the environment perspective, what are crucial policy interventions from government to ensure water supply sector in Malaysia is sustainable?
11. Do you think the regulator (NWSC) be able to enforce WSIA 2006 effectively, including giving due consideration to balanced development between environmental and economic considerations?
12. If no, what prohibits NWSC from discharging its functions effectively? If yes, what have NWSC done with regard to environment protection?

13. Please suggest how could environmental issues /protection be given greater consideration in the sector?
14. Please suggest how can environmental organizations (like WWP/WWF/MNS) play important roles in the water sector?
15. How can they be implemented?

## D. *Government officials*

1. What do you envisage to achieve from that reform?
2. Have they been achieved so far?
3. Do you satisfy with the performance/progress (of the reform) thus far?
4. If no, why? If yes, in what areas do you satisfy the most?
5. What are your top most concerns about the reform?
6. Do your concerns being addressed in the reform?
7. If no, what areas are being left out of the reform?
8. Please suggest how can they being included in the reform?
9. Were you consulted during the reform? If no, why?
10. If yes, what have you proposed?
11. Did you manage to convince the government during the reform process?
12. Did you act alone, or in coalition with others?
13. If no, why? If yes, who were they? Was that coalition effectives? If no, why?
14. Do you think NWSC can enforce WSIA 2006 effectively to bring about sustainable water supply sector in Malaysia?
15. If no, what prohibits NWSC from discharging its responsibility effectively?
16. If yes, in what areas did NWSC perform effectively?
17. Do you satisfy with the performance NWSC thus far?
18. If no, why? If yes, please state in what areas?
19. What are needed to ensure NWSC remains impartial in discharging it duties?
20. How can NWSC or the government facilitate efficient and effective water supply management in Malaysia?
21. Do you think certain provisions in WSIA 2006 have to be amended to bring about greater efficiency and effectiveness to the sector?
22. If no, why? If yes, what amendments do you suggest? How can they help to improve the overall sector?
23. What are formulas or ingredients for NWSC to carry out its responsibility as independence regulator?
24. How can we ensure NWSC strike a balance between safeguarding the interests of industry (operators) and consumers?
25. What are crucial policy interventions from government to ensure the sustainability of the water supply sector in Malaysia?
26. Do you see the involvement of all stakeholders in the sector (consumers, environmentalists, NGOs, civil society) are vital to the sustainability of the sector?

27. If no, why? If yes, please suggest how other stakeholders (consumers, environmentalists, NGOs, civil society) could contribute to enhance sustainability in the sector?
28. What measures do you want to propose to be included in the reform in the future?
29. How can they be implemented?
30. What is the position of the government in response to the request from MWA and water utilities to exempt sludge as Schedule Waste under EQA 1974?
31. Do you think water supply sector should remain in public hands rather than being privatised?
32. What is your opinion about the suggestion for the government to stop furhter water privatisation?
33. What will be the best model for efficient, effective and sustainable Malaysia water industry?
34. What are measures crucial to be included in the reform in the future?
35. How can they be implemented?

## Appendix 2. Organization structure of NWSC 2011

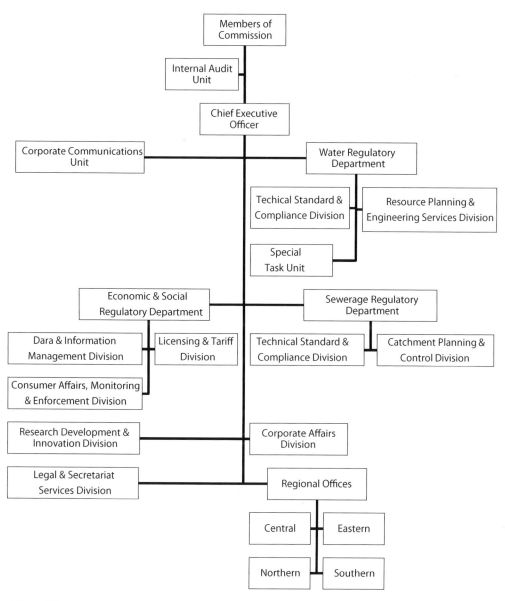

*Adapted from www.span.gov.my.*

## Appendix 3. Organization structure of PAAB 2012

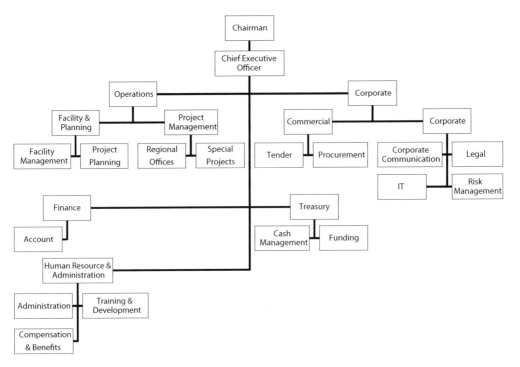

*Adapted from www.paab.my.*

## Appendix 4. Domestic and industrial water tariffs in Peninsular Malaysia (2009)[a]

| State | Water utilities | RM sen/m³ (domestic)[b] | Ranking (lowest to highest) |
|---|---|---|---|
| Penang | PBAPP | 0.31 | 1 |
| Terengganu | SATU | 0.52 | 2 |
| Kedah | SADA | 0.53 | 3 |
| Kelantan | AKSB | 0.55 | 4 |
| Pahang | JBA Pahang | 0.57 | 5 |
| Perlis | PWD Perlis | 0.57 | 5 |
| N. Sembilan | SAINS | 0.68 | 6 |
| Melaka | SAMB | 0.72 | 7 |
| Perak | LAP | 0.73 | 8 |
| Selangor | Syabas | 0.77 | 9 |
| Johor | SAJH | 0.98 | 10 |

| State | Water utilities | RM sen/m³ (industrial)[c] | Ranking (lowest to highest) |
|---|---|---|---|
| Terengganu | SATU | 1.15 | 1 |
| Penang | PBAPP | 1.19[d] | 2 |
| Kedah | SADA | 1.20 | 3 |
| Kelantan | PWD Perlis | 1.25 | 4 |
| Perlis | JBA Pahang | 1.30 | 5 |
| Pahang | JBA Pahang | 1.45 | 6 |
| Melaka | SAMB | 1.47 | 7 |
| N. Sembilan | SAINS | 1.59 | 8 |
| Perak | LAP | 1.60 | 9 |
| Selangor | Syabas | 2.27 | 10 |
| Johor | SAJH | 2.93 | 11 |

[a] Source: MWA, 2010.

[b] For first 300 m³.

[c] For first 500 m³.

[d] After 1.11.2010 review.

## Appendix 5. List of companies under PBAHB

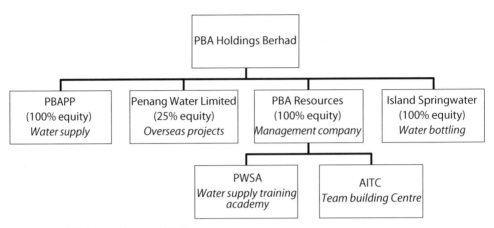

*Source: PBAHB Annual Report, 2009.*

## Appendix 6. SADA's water tariffs in comparison with other utilities[a]

| State | Water utilities | 2009 (RM) | | | Ranking (lowest to highest) |
|---|---|---|---|---|---|
| | | Domestic[b] | Trade[c] | Overall | |
| Penang | PBAPP | 0.31 | 1.19 | 0.70 | 1 |
| Terengganu | SATU | 0.52 | 1.15 | 0.84 | 2 |
| Kedah | SADA | 0.53 | 1.2 | 0.87[d] | 3 |
| Kelantan | AKSB | 0.55 | 1.25 | 0.90 | 4 |
| Labuan | JBA Labuan | 0.90 | 0.90 | 0.90 | 5 |
| Perlis | PWD Perlis | 0.57 | 1.30 | 0.94 | 6 |
| Pahang | JBA Pahang | 0.57 | 1.45 | 1.01 | 7 |
| Melaka | SAMB | 0.72 | 1.47 | 1.10 | 8 |
| N. Sembilan | SAINS | 0.68 | 1.59 | 1.14 | 9 |
| Perak | LAP | 0.73 | 1.60 | 1.17 | 10 |
| Selangor | Syabas | 0.77 | 2.27 | 1.52 | 11 |
| Johor | SAJH | 0.98 | 2.93 | 1.96 | 12 |
| Malaysia | NA[e] | 0.65 | 1.32 | 0.99 | NA |

[a] Source: MWA, 2010.

[b] For first 35 m$^3$.

[c] For first 500 m$^3$.

[d] Before tariff rationalization.

[e] NA: not applicable.

## Appendix 7. Basic flow of complaint management for SADA

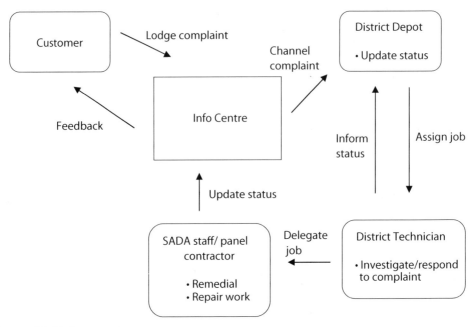

*Source: SADA document.*

## Appendix 8. SADA's e-complaint form

*Source: www.sada.com.my.*

## Appendix 9. List of interviews

| Interviewees/designation | Organization | Date |
|---|---|---|
| **Water companies** | | |
| Mr. Wan Hamdy Wan Ibrahim<br>Executive Director | Air Utara Indah Sdn. Bhd | 6.5.09 |
| Mr. Mohd. Norazi Mohd Nordin<br>Assistant General Manager | Air Utara Indah Sdn. Bhd | 6.5.09 |
| Ms. Annie Chai Ai Nai<br>General Manager | Salcon Engineering Berhad | 25.5.09 |
| Mr. Ir. Abas Abdullah<br>Chief Operating Officer | Konsortium ABASS | 19.6.09 |
| Ir. Ainul Azhar Mohd. Jamoner<br>Assistant General Manager | Konsortium ABASS | 19.6.09 |
| Mr. Abd. Hamid bin Sahid<br>Senior Water Engineer | Perlis Water Supply Department | 28.4.09 |
| Madam Noriah Bt Ismail<br>Distribution Manager | Syarikat Air Terengganu Berhad | 6.7.09 |
| Ir. Mohd Yusof Mohd Isa<br>General Manager | Perak Water Board | 22.5.09 |
| Mr. Ishak Abd. Rahman<br>Senior Engineer | Perak Water Board | 22.5.09 |
| Mr. Harun Jasin<br>Director | Kedah Water Supply Department | 11.5.09 |
| Mr. Zulkepli Mishat<br>General Manager | GSL Water | 13.5.09 |
| Ir. Ng Seik Long<br>Group General Manager (Water and Engineering<br>Division) | Taliworks Corporation Berhad | 8.5.09 |
| Mr. Wong Mean<br>Group General Manager (Business Development) | Taliworks Corporation Berhad | 8.5.09 |
| Mr. Khairudin Bin Din<br>Controller Padang Saga WTP | Taliworks Corportion Berhad | 9.8.10 |
| Ir. Ismail Mohd. Zain<br>General Manager (Technical) | Syarikat Air Darul Aman | 6.8.10 |
| Ir. Shahruddin Othman<br>Special officer | Syarikat Air Darul Aman | 19.7.10 |
| Madam Siti Syuhairah Sobri<br>Customer Service Manager | Syarikat Air Darul Aman | 9.3.11 |
| Mr. Ahmad Fauzi Abdul Rahman<br>Information Center Supervisor | Syarikat Air Darul Aman | 1.8.10 |

| Interviewees/designation | Organization | Date |
|---|---|---|
| Madam Salina Ismail<br>Manager (Legal) | Syarikat Air Darul Aman | 8.8.10 |
| Ir. Ong Eng Chuan<br>Production Manager | Perbadanan Bekalan Air Pulau Pinang | 2.11.09 |
| Ir Jaseni Maidinsa<br>General Manager | Perbadanan Bekalan Air Pulau Pinang | 4.11.09 |
| Ir. Kan Cheong Weng<br>Strategic Planning Manger | Perbadanan Bekalan Air Pulau Pinang | 12.11.09 |
| Madam Khairulbariah Dzun-Nurin<br>Customer Service Manager | Perbadanan Bekalan Air Pulau Pinang | 26.10.09 |
| Madam Noridah Abd Kadir<br>Perai CCC Supervisor | Perbadanan Bekalan Air Pulau Pinang | 3.11.09 |
| Ir. K. Jeyabalan<br>Manager (Corporate Affairs) | Perbadanan Bekalan Air Pulau Pinang | 7.7.09 |
| Mr. Zakaria Mohammad Sultan<br>Senior Executive (Legal) | Perbadanan Bekalan Air Pulau Pinang | 7.7.09 |
| Mr. Sani Sidik<br>Senior General Manager Group Business Development | Metropolitan Utilities Corporation Sdn. Bhd. | 4.5.09 |
| Mr. Patrick Sim Siew Min<br>General Manager (Operation) | Gamuda Water Sdn. Bhd. | 27.5.09 |
| Ms. Chew chiew Ean<br>Senior Manager (Finance) | Gamuda Water Sdn. Bhd. | 27.5.09 |
| Dato' Matlasa Hitam<br>Managing Director | Puncak Niaga Sdn. Bhd | 18.6.09 |
| Mr. Ir. V. Subramaniam<br>Executive Director (Operations) | Syarikat Pengeluar Air Selangor Sdn. Bhd. | 23.7.09 |
| Mohamad Hairi Basri<br>Secretary-General | Malaysian Water Association | 3.3.11 |
| Mr. Ahmad Zahdi Jamil<br>Chief Executive Officer | Syarikat Air Johor Holdings | 30.4.09 |
| Mr. David CS Lim<br>Director | Aliran Ehsan Resources Berhad | 21.5.09 |
| Mr. Ismail Mat Nor<br>Director | Pahang Water Supply Department | 14.5.09 |
| Mr. Mohd. Ashri Awang<br>Manager (Operation) | Air Kelantan Sdn. Bhd. | 3.6.09 |
| Ir. Zulkipli Ibrahim<br>General Manager | Negeri Sembilan Water Company | 5.5.09 |
| Ir. Mohd. Khalid Nasir<br>Chief Executive Officer | Syarikat Air Melaka Berhad | 25.6.09 |

| Interviewees/designation | Organization | Date |
|---|---|---|
| **Consumer associations** | | |
| Mr. Piarapakaran S. Chief Operating Officer | Federation of Malaysian Consumers Associations | 10.8.09 |
| Ms. Mageswari Sangaralingam Research Officer | Consumers' Association of Penang | 1.9.09 |
| Mr. K. Koris President | Penang Consumer Protection Association | 3.8.09 |
| Mr. Mohidden Abdul Kader Vice President | Consumers' Association of Penang | 1.9.09 |
| **Environmental organizations** | | |
| Dr. Loh Chi Leong Executive Director | Malaysian Nature Society | 11.8.09 |
| Mr. Abdul Razak Lubis Researcher | Water Watch Penang | 14.8.09 |
| Prof. Dr. Chan Ngai Weng President | Water Watch Penang | 5.8.09 |
| Dato' Prof. Dr. Anwar Fazal Advisor | Water Watch Penang | 5.8.09 |
| **Government officials** | | |
| Mr. Sutekno Ahmadbelon Under-Secretary Water Services Regulation Division | Ministry of Energy, Water and Communications | 17.9.09 |
| Ir. Lee Koon Yew Executive Director | National Water Services Commission | 9.9.09 |
| Ms Leow Peen Fong Regulatory Director | National Water Services Commission | 15.9.09 |
| Dato' Teo Yen Hua Chief Executive Officer | National Water Services Commission | 23.9.09 |
| Mr. Hashim Daud Director, Water and Marine Division | Department of Environment | 2.10.09 |
| Ir. Lee Heng Keng Deputy Director-General | Department of Environment | 16.10.09 |
| Mr. Mohd. Nazry Mohd Kassim Principal Assistant Director | Economic Planning Unit, Prime Minister's Department | 14.10.09 |
| Ir. Amiruddin Hamzah State Minister for Water Supply | Kedah state government | 19.7.10 |
| Ir. Noor Azahari Zainal Abidin Director Planning, Coordinating and Monitoring Division | Water Supply Department, Ministry of Energy, Green Technology and Water, Malaysia | 4.3.11 |

## Appendix 10. Survey questionnaires

Dear valued respondent

I was formerly an employee of the then Ministry of Energy, Water and Communications, Malaysia from 2004-2008. In 2008, I was granted a study leave and at the same time was offered a scholarship by Public Service Department of Malaysia to further study in the area of water management. Currently, I am pursuing my Ph.D program at Wageningen University in the Netherlands. My research is entitled 'Operational Efficiency and Environmental Effectiveness of the Malaysian Water Supply Sector'. Attached to this is also a letter pertaining to the research from my supervisor. My research encompasses an extensive analysis of the performance of all water operators in Malaysia, particularly before and after the restructuring of water supply sector which was initiated in 2006.

Your kind cooperation in giving me and/or my research assistants an interview, and to fill in the attached questionnaire is highly appreciated. I would appreciate it very much if you could kindly fill in the attached questionnaire and remit it in the enclosed stamped self-addressed envelope to:

Mr. Ching Thoo a/I Kim
7B-05-06, Tiara Intan Kondo
Jalan Bukit Indah 3/19
Taman Bukit Indah
68000 Ampang
Selangor
Email: ching.kim@wur.nl
Contact no: 019-7722375

Your kind cooperation is highly appreciated.

Yours sincerely

Ching Thoo a/1 Kim

Date: 24 April 2009

# General instruction to respondents

1. Please fill in the information about your organization, its address, and the date.
2. Please tick the box that is closest to your opinion.
3. If you hesitate, please tick the box that first comes to your mind, as this mostly represents your closest opinion.

## I. *Treatment only operators*

Your organization:
Address:
Date:

## A. *General information*

1. How many treatment plants do you manage/own?
   - <3
   - 3-6
   - 7-9
   - More than 9 (please state)

2. What are the capacities of those treatment plants?
   - <250 mgd
   - 250-500 mgd
   - 501-750 mgd
   - 751-1000 mgd
   - More than 1000 mgd (please state)

3. What are the average ages of those treatment plants?
   - <3 years
   - 3-5 years
   - 6-8 years
   - 9-11 years
   - More than 12 years

4. By which means were that treatment plants built/operated?
   - Management contract (O&M)
   - Lease
   - BOT/BOOT
   - Concession
   - Divestiture
   - Other (please state)

5. If you were to operate/manage the treatment plants by means mentioned in Question 4, please state the duration of such contracts?

| Mode of operations | Duration of contract (years) |
|---|---|
| Management contract (O&M) | |
| Lease | |
| BOT/BOOT | |
| Concession | |
| Divestiture | |
| Other (Please state) | |

## B. Sludge treatment facilities

6. In all the water treatment plants you manage, please state the following:

| # of WTP equipped with sludge treatment facilities | # of WTP without sludge treatment facilities |
|---|---|
| | |

7. For those treatment plants without sludge treatment facilities:
   a. Please state where does the sludge is being disposed of?
   b. Why do those treatment plants not have sludge treatment facilities?
   c. Are you taking any initiatives to build on-site sludge treatment facilities for those treatment plants?
   - No (please answer Question 8-9)
   - Yes

8. What are the reasons for not building sludge treatment facilities for those treatment plants? Please state your reasons below.

9. Are you sending the sludge for treatment at private sludge treatment facilities?
   - No (please answer Question 10
   - Yes

10. If no, please explain why?

For treatment plants equipped with sludge treatment facilities:

11. Please state technology/facilities used in those plants.

12. What are you doing with the re-cycled sludge?
   - Disposed of at landfill
   - Convert them for other usage (please state)
   - Others (please state)

13. Are you taking any initiatives to use the re-cycled sludge/converting sludge for other usage?
   - No (please answer Question 14)
   - Yes (please answer Question 15-16)

14. What prevents you from not converting re-cycled sludge for other usage?
   - They do not have value
   - Costly
   - No demand
   - No incentive from government
   - Others (please state)

If yes:

15. What new usage do you generate from re-cycled sludge? Please state them.

16. In your opinion, what must be done to encourage/facilitate the usage of re-cycled sludge among water operators?

## C. Conformance to Environment Quality Act 1974

17. Are you aware of regulation(s) under Environment Quality Act (EQA) 1974 about sludge treatments?
   - No
   - Yes (please answer Question 18)

18. Please name/state regulations under EQA 1974 regarding sludge treatments that you know of?

19. Does sludge disposal from your treatment plants conform to the regulation(s) under EQA 1974 that you have stated in Question 18?
   - No (please answer Question 20-21)
   - Yes (please answer Question 22)
   - Don't know

20. What are the reasons for not conforming to the sludge disposal as required by that regulation(s) under EQA 1974?
   - They were old plants build without sludge treatment facilities
   - Costly to upgrade to meet the EQA 1974
   - Lack of enforcement from DOE
   - Others (please state)

21. Are you taking any action to conform to the EQA 1974?
    - No
    - If yes, what are they? Please list them below.

If yes:

22. Please provide number of treatment plants that meet the sludge disposal standard of that regulation (s) under EQA 1974?

| # of WTP managed | # of WTP that conform to EQA 1974 for sludge disposal standard |
|------------------|-------------------------------------------------------------|
|                  |                                                             |

23. Do you think the EQA 1974 is effective to handle sludge treatment?
    - No (please answer Question 24)
    - Yes (please answer Question 25-26)
    - Don't know

24. Why EQA 1974 is not effective to handle sludge treatment?
    - It's comprehensive regulation, but poor enforcement
    - Fines/penalties are too low
    - Lack coordination among related government agencies
    - Others (please state)

If yes:

25. In what way do you think that EQA 1974 is an effective regulation with regard to the sludge treatment?

26. In your opinion, how the implementation of EQA 1974 could be improved?

27. Do you have problems conforming to EQA 1974?
    - No
    - Yes (please answer Question 28)

28. If yes, why? Please state the reasons.

29. What are the non-conformance rates for the following years?

| Year | No. of plants managed | No. of plants didn't conform to EQA 1974 for sludge treatment |
|------|----------------------|---------------------------------------------------------------|
| 2005 |                      |                                                               |
| 2006 |                      |                                                               |
| 2007 |                      |                                                               |
| 2008 |                      |                                                               |
| 2009 |                      |                                                               |

30. In your opinion, what must Department of Environment (DOE) do to facilitate/promote/encourage sludge treatment among water treatment operators in Malaysia?
    - Beef up the enforcement of EQA 1974
    - Increase the fines/penalties
    - Provide incentives i.e. tax exemption, levy for acquiring new technologies and converting re-cycled sludge for new usage, etc.
    - Implement 'polluters pay' principle
    - Others (please state)

31. Do you think it is timely to incorporate environment costs into water tariffs?
    - No (please answer Question 32)
    - Yes (please answer Question 33-34)

32. If no, state your reasons.

33. If yes, why?

34. How can it be implemented?

35. In your opinion, besides EQA 1974, what other aspects of environmental performance/regulations should government takes into consideration? Please list your suggestions.

36. Please name awards or accreditations received for environmental management?

| Year | Awards/accreditation received | Awarded by |
|------|------------------------------|------------|
|      |                              |            |
|      |                              |            |
|      |                              |            |
|      |                              |            |
|      |                              |            |
|      |                              |            |

## D. Drinking water compliance standard

37. Where do you get raw water for treatment?

38. If you get the supply of raw water from rivers, lakes, canals, what is the state/quality of those catchment areas? (According to DOE's classification).
   - Clean
   - Slightly polluted
   - Polluted
   - Very polluted

39. Are you being charged by state governments for raw water abstraction?
   - No
   - Yes, how much are you being charged for (cent/m$^3$)

40. Since water catchments fall under jurisdiction of State, do you think State is doing enough to conserve rivers/look after raw water quality?
   - No (please answer Question 41)
   - Yes (please answer Question 42)

41. Why do you think that State is not doing enough to conserve rivers/look after raw water quality?
   - Lack of funds/allocations
   - Lack of human resources
   - Lack of expertise
   - Lack cooperation from Federal agencies
   - Others (please state)

42. Despite efforts by State, what other measures that State could initiate to conserve rivers/look after raw water quality? Please list your suggestions.

43. Do you experience raw water quality problems?
   - No
   - If yes, what do you do? Please list your actions

44. Do you have treated water labs?
   - No
   - Yes (please answer Question 45)

If yes:

45. How many chemists do you employed to man the lab?

46. Do you send water sampling to Chemistry Department for testing/validation?
    - No
    - If yes, please state the frequency

47. Do you collaborate with Chemistry Department to ensure conformance to National Guideline for Drinking Water Quality Standard(NGDWS)?
    - No (please answer Question 48)
    - Yes (please answer Question 49)

48. What are the reasons for non-existence of such cooperation? Please state the reasons.

If yes:

49. State how you benefit from collaboration with Chemistry Department with regard to the drinking water quality.

50. In what forms does the collaboration exist?
    - Scheduled meeting (weekly/monthly, etc.)
    - Changing of information (reports/documents, etc.)
    - Workshops/seminars
    - Others (please state)

51. In the event of non-conformance to NGDWS, what remedial actions do you initiate?
    - Review raw water treatment processes
    - Cease treatment for the affected plants
    - Re-examined raw water quality at intake points
    - Liaise/cooperate closely with Chemistry Department to determine root causes
    - Others (please state)

52. In the past five years, what are the incidences of non-compliance to NGDWS occurred?

| Year | No. of non-compliances with NGDWS |
|------|-----------------------------------|
| 2005 | |
| 2006 | |
| 2007 | |
| 2008 | |
| 2009 | |

53. What plans do you have to improve the compliance rate to NGDWQ? Please state.

54. Besides NGDWQ, what are other aspects equally feasible to measure drinking water compliance among water operators? Please list your suggestions.

55. Please name awards or accreditations received for drinking water quality management?

| Year | Awards/accreditation received | Awarded by |
|------|-------------------------------|------------|
|      |                               |            |
|      |                               |            |
|      |                               |            |
|      |                               |            |
|      |                               |            |
|      |                               |            |

## E. Environmentally-friendly treatment process

56. Please provide number of plants that used environmental-friendly versus conventional treatment process?

| Year | No. of plants managed | No. of plants that used conventional treatment process | No. of plants that used environmental-friendly treatment process |
|------|-----------------------|--------------------------------------------------------|-------------------------------------------------------------------|
| 2005 |                       |                                                        |                                                                   |
| 2006 |                       |                                                        |                                                                   |
| 2007 |                       |                                                        |                                                                   |
| 2008 |                       |                                                        |                                                                   |
| 2009 |                       |                                                        |                                                                   |

57. Why are the reasons for not switching to environmental-friendly treatment process for all plants?

58. When do you plan to convert all plants to environmental-friendly treatment process?
- 1 year from now
- 2 years from now
- 3 years from now
- Don't know yet

59. Please list what type of environmental-friendly treatment processes/technologies do you intend to use?

## F. Unit production cost

60. In average, what is the cost to produce one cubic meter of treated water for last 5 years?

| Year | Production cost (RM/m$^3$) |
|------|----------------------------|
| 2005 |                            |
| 2006 |                            |
| 2007 |                            |
| 2008 |                            |
| 2009 |                            |

61. What are the percentages of the following cost for treatment of water?

| Element | % |
|---------|---|
| Energy |  |
| Chemical |  |
| Labour |  |
| Distribution |  |
| Financing/loan servicing costs |  |

62. Do you initiate cost cutting measures?
   - No (please answer Question 63)
   - Yes (please answer Question 64-65)

63. If no, why?

64. If yes, which cost elements do you emphasis? Please rank them in order from 1-5 (1 being the top most priority, and 5 being the least priority)

| Element | Rank (1-5) |
|---------|------------|
| Energy |  |
| Chemical |  |
| Labour |  |
| Distribution |  |
| Financing/loan servicing costs |  |

65. To what degree, do the cost cutting measures help the company?
   - Help in lowering overall costs
   - Achieve higher profits
   - Better remuneration scheme to employees
   - Others (please state)

## G.  *Information disclosure on sludge management (EQA 1974)*

66. Do you record compliance to EQA 1974 for sludge disposal standards?
    - No (please answer Question 67)
    - Yes (please answer Question 68-69)

67. If no, why? Please state reasons:

If yes:

68. How do you record them?
    - By individual treatment plants
    - Overall compliance
    - Others (please state)

69. In what forms are they recorded?
    - Digital forms
    - Non-digital forms

70. Do you publish compliance to EQA 1974?
    - No (please answer Question 72)
    - Yes (please answer Question 73-74)

71. If no, what prevent you from not publishing them?

If yes:

72. How do you publish them?
    - By individual treatment plants
    - Overall compliance
    - Others (please state)

73. In what forms/mediums are they published?
    - Monthly/annual reports
    - Company's leaflets
    - Press releases/statements
    - Published reports
    - Company's websites
    - Others (please state)

74. Do you have/implemented environmental policy (policy statement, pledge, charter)?
    - No (please answer Question 75)
    - Yes (please answer Question 76)

75. If no, when do you plan to implement them?
    - 1 year from now
    - 2 years from now
    - 3 years from now
    - Don't know yet

76. If yes, since when do you implement them?
    - 1-3 years ago
    - 4-6 years ago
    - 7-9 years ago
    - More than 10 years ago

77. Do you also implement Environmental Management System (EMS) such as ISO 14001?
    - No (please answer Question 78)
    - Yes (please answer Question 79)

78. If no, when do plan to implement them?
    - 1 year from now
    - 2 years from now
    - 3 years from now
    - Don't know yet

79. If yes, since when do you implement them?
    - 1-3 years ago
    - 4-6 years ago
    - 7-9 years ago
    - More than 10 years ago

## H. *Information disclosure on drinking water conformance standard*

80. Do you record compliance to National Guideline for Water Drinking Quality Standard (NGWDS) issues by Ministry of Health for each of the treatment plants?
    - No (please answer Question 81)
    - Yes (please answer Question 82-83)

81. If no, why?

If yes:

82. How do you record them?
    - By individual treatment plants
    - Overall compliance
    - Others (please state)

83. How do you keep the compliance records?
- Digital forms
- Non-digital forms

84. Do you publish compliance to NGDWS?
- No (please answer Question 85)
- Yes (please answer Question 86-87)

85. If no, what prevent you from not publishing them?

If yes:

86. How do you publish them?
- By individual treatment plants
- Overall compliance
- Others (please state)

87. In what forms/mediums are they published?
- Monthly/annual reports
- Company's leaflets
- Press releases/statements
- Published reports
- Company's websites
- Others (Please state)

88. Do you have/implemented drinking water quality policy (policy statement, pledge, charter)?
- No (please answer Question 89)
- Yes (please answer Question 90)

89. If no, when do you plan to implement them?
- 1 year from now
- 2 years from now
- 3 years from now
- Don't know yet

90. If yes, since when do you have them?
- 1-3 years ago
- 4-6 years ago
- 7-9 years ago
- More than 10 years ago

## I. Information disclosure under Section 29 of Water Services Industry Act (WSIA) 2006

91. Are you aware that information disclosure is mandatory under WSIA 2006?
    - No
    - Yes (please answer Question 92-94)

If yes:

92. What steps does your company undertake to comply with information disclosure (of environmental performance and drinking water quality compliance) under WSIA 2006? Please list them.

93. What problems do you envisage in order to comply with this requirement? Please list them.

94. What do you think the National Water Services Commission (NWSC) should do to encourage/ facilitate information disclosure among water operators? Please state your suggestions:

95. Do you think NWSC can enforce WSIA 2006 effectively in promoting information disclosure among water operators?
    - No (please answer Question 96)
    - Yes (please answer Question 97-98)

96. If no, please state your reasons.

If yes:

97. What do you say so?

98. As central regulator for water industry in Malaysia, what assistance does NWSC require to undertake this task (information disclosure) in a more effective manner?

99. Some say that information disclosure (documenting and publishing conformance/compliance environmental performance) will enhance the image of water operators, thus, reflects viable business strategies? What do you think of this statement?
    - Very likely
    - Likely
    - Not sure
    - Unlikely
    - Very unlikely

100. Based on your opinion in Question 99 above, why do you think so?

101. In what way publishing environmental performance could enhance the image of water operators as business entities in the long run?

## II. Distribution only operators

### A. Non-revenue water (NRW)

1. Please state NRW performance for the last 5 year (in %).

| Component | 2005 | 2006 | 2007 | 2008 | 2009 |
|---|---|---|---|---|---|
| Real losses | | | | | |
| Apparent losses | | | | | |
| Total | | | | | |

2. Are you instituting globally-accepted methodology in calculating NRW within your organization?
   - No (please answer Question 3)
   - Yes (please answer Question 4-5)

3. If no, please state reasons.

If yes:

4. What are they? Please state them.

5. Why do you choose to use them?

6. Please state your allocation for NRW reduction programmes for last 5 years?

| Year | Amount (RM) million |
|---|---|
| 2005 | |
| 2006 | |
| 2007 | |
| 2008 | |
| 2009 | |

7. How do you fund that programmes?
   - Federally funded
   - State funding
   - Private borrowing
   - Others (please state)

8.  Please list 5 major programmes that have been implemented to reduce NRW in your organization?

9.  Based on 5 programmes you have listed in Question 8, which programmes yield the most effective results? Please rank them from 1 (least effective) – 5 (most effective).

| Programme | Rank (1-5) |
|---|---|
| 1. | |
| 2. | |
| 3. | |
| 4. | |
| 5. | |

10. Do you learn/benchmarking your NRW performance with other operators?
    - No (Please answer Question 11)
    - Yes (Please answer Question 12-13)

11. If no, please state your reasons.

If yes:

12. Who are they?

13. Please state what have you learned from those operators?

14. In your opinion, what could water operators initiate to improve NRW performance to an acceptable level? Please list them.

15. In what ways do you think the restructuring of water supply industry is able to assist water operators to improve their NRW performance in general? Please list your opinions.

16. What policy interventions by government/regulator are needed to drive/facilitate water operators to achieve at acceptable NRW level? Please list them.

## B. Collection rate efficiency

17. How do your consumers pay the bills?

18. How many consumers' accounts do you manage in total?

| Category | # accounts as of now |
|---|---|
| Domestic | |

| Non-domestic | |
|---|---|
| Total | |

19. In last five years, please state your collection rate efficiency?

| Year | No. of bills issued | No. of bills collected | % bills collected |
|---|---|---|---|
| 2005 | | | |
| 2006 | | | |
| 2007 | | | |
| 2008 | | | |
| 2009 | | | |

20. Are you happy with the current performance (of collection rate efficiency)?
    - If no (please answer Question 21)
    - If yes

21. If no, please state what have you done to improve the performance to a satisfactory level?

22. Do you learn from /benchmark your collection rate efficiency performance with other operators?
    - No (please answer Question 23)
    - Yes (please answer Question 24-25)

23. If no, please state your reasons:

If yes:

24. Who are they?

25. Please state what have you learned from those operators?

26. In your opinions, what water operators could initiate the improvement of collection rate efficiency? Please list them.

27. Do you think the restructuring of water supply industry can lead to collection rate efficiency in general? If so, how? Please list your opinions.

28. What policy interventions by government/regulator are needed to facilitate water operators to improve collection rate efficiency? Please list them.

## C. *Customer service complaints*

29. Do you have customer service centre?
    - No (please answer Question 30-31)
    - Yes (please answer Question 32-34)

If no:

30. What were the reasons for not having customer service centre?

31. Do you plan to have them in the future?
    - No
    - Yes (please state when do you want to establish them)

If yes:

32. Where are they located?

33. Number of personnel deployed to man the centre?

34. Type of customer service technology/software used at centre?

35. Please state no. of complaints received over the last five years (service and technical related complaints)

| Year | # service related complaints | # technical related complaints | Total |
|------|------------------------------|---------------------------------|-------|
| 2005 | | | |
| 2006 | | | |
| 2007 | | | |
| 2008 | | | |
| 2009 | | | |

36. Please list 5 most common complaints for both categories?

| Category | Five most common complaints |
|---|---|
| Service related complaints | 1. |
| | 2. |
| | 3. |
| | 4. |
| | 5. |
| Technical related complaints | 1. |
| | 2. |
| | 3. |
| | 4. |
| | 5. |

37. Please explain what do you do to address these complaints?

38. How long do you normally (in average) take to address these complaints?

39. Do you learn from /benchmark your customer service performance with other operators?
   - No (please answer Question 40)
   - Yes (please answer Question 41-42)

40. If no, please state your reasons:

If yes:

41. Who are they?

42. Please state what have you learned from those operators?

43. In your opinion, what water operators could initiate the improvement of customer service complaints? Please list them.

44. In what ways do you think the restructuring of water supply industry is able to assist water operators to improve customer service complaints in general? Please list your opinions.

45. What policy interventions by government/regulator are needed to facilitate water operators to improve customer service? Please list them.

46. In your opinion, what would a customer service centre mean to water operators? Please list them.

47. In what ways customer service centre helps to enhance image of water operators?

## D. Drinking water compliance standard

48. Does treated water supplied by treatment operators conforms to NGDWS standard?
    - No (please answer Question 49)
    - Yes

49. Please state your action to rectify the problem?

50. Do you send water sampling to Chemistry Department for testing/validation?
    - No
    - If yes, please state the frequency)

51. Do you collaborate with Chemistry Department to ensure conformance to National Guideline for Drinking Water Quality Standard (NGDWS)?
    - No (please answer Question 52)
    - Yes (please answer Question 53-54)

52. What are the reasons for non-existence of such cooperation? Please state the reasons:

If yes:

53. State how you benefit from collaboration with Chemistry Department with regard to the drinking water quality:

54. In what forms does the collaboration exist?
    - Scheduled meeting (weekly/monthly, etc.)
    - Changing of information (reports/documents, etc.)
    - Workshops/seminars
    - Others (please state)

55. In the event of non-conformance to NGDWS, what remedial actions do you initiate?
    - Review raw water treatment processes
    - Cease treatment for the affected plants
    - Re-examined raw water quality at intake points
    - Liaise/cooperate closely with Chemistry Department to determine root causes
    - Others (please state)

56. In the past 5 years, what are the incidences of non-compliance to NGDWS occurred?

| Year | No. of non-compliances to NGDWS 2001 |
|------|--------------------------------------|
| 2005 | |
| 2006 | |
| 2007 | |
| 2008 | |
| 2009 | |

57. What plans do you have to improve the compliance rate to NGDWS? Please state:

58. Besides NGDWS, what are other aspects equally feasible to measure drinking water compliance among water operators? Please list your suggestions

## E. *Information disclosure on drinking water conformance standard*

59. Do you record compliance to National Guideline for Water Drinking Quality Standard (NGWDS) issues by Ministry of Health?
   - No (please answer Question 60)
   - Yes (please answer Question 61-62)

60. If no, why?

If yes:

61. Please state how you record them?

62. How do you keep the compliance records?
   - Digital forms
   - Non-digital forms

63. Do you publish compliance to NGDWS?
   - No (please answer Question 64)
   - Yes (please answer Question 65-66)

64. If no, what prevent you from not publishing them?

If yes:

65. Please state how do you publish them?

66. In what forms/mediums are they published?
    - Monthly/annual reports
    - Company's leaflets
    - Press releases/statements
    - Published reports
    - Company's websites
    - Others (please state)

67. Do you have/implemented drinking water quality policy (policy statement, pledge, charter)?
    - No (please answer Question 68)
    - Yes (please answer Question 69)

68. If no, when do you plan to implement them?
    - 1 year from now
    - 2 years from now
    - 3 years from now
    - Don't know yet

69. If yes, since when do you have them?
    - 1-3 years ago
    - 4-6 years ago
    - 7-9 years ago
    - More than 10 years ago

## F.  Information disclosure under Water Services Industry Act (WSIA) 2006

70. Are you aware that information disclosure is mandatory under WSIA 2006?
    - No
    - Yes (please answer Question 71-72)

If yes:

71. What steps does your company undertake to comply with information disclosure (of environmental performance and drinking water quality compliance) under WSIA 2006? Please list them.

72. What problems do you envisage in order to comply with this requirement? Please list them.

73. What do you think National Water Services Commission (NWSC) should do to encourage/ facilitate information disclosure among water operators? Please state your suggestions:

74. Do you think NWSC can enforce WSIA 2006 effectively in promoting information disclosure among water operators?
    - No (please answer Question 75)
    - Yes (please answer Question 76-77)

75. If no, please state your reasons.

If yes:

76. What do you say so?

77. As central regulator for water industry in Malaysia, what assistance does NWSC require to undertake this task (information disclosure) in a more effective manner?

78. Some say that information disclosure (documenting and publishing conformance/compliance environmental performance) will enhance the image of water operators, thus, reflects viable business strategies? What do you think of this statement?
    - Very likely
    - Likely
    - Not sure
    - Unlikely
    - Very unlikely

79. Based on your opinion in Question 78 above, why do you think so?

80. In what way publishing environmental performance could enhance the image of water operators as business entities in the long run?

## Appendix 11. Compliance to drinking water quality standard

| Category | Water utilities | # violation detected |
|---|---|---|
| Fully compliance | Taliworks, Salcon Engineering, Equiventures, Konsortium ABASS, PBAPP | No violations detected |
| Violations detected | Gamuda Water | 2005-2007: 0<br>2008: 5 |
| | Syabas[1] | 2005: 2.47% non-conformance<br>2006: 2.28% non-conformance<br>2007: 1.11% non-conformance<br>2008: 0.58% non-conformance |
| | GSL Water | 2005: 12<br>2006: 2<br>2007: 9<br>2008: 2 |
| | Air Utara Indah | 2005: 14<br>2006: 18<br>2007: 17<br>2008: 12 |
| | AKSB | Turbidity: no. of violations(s) not indicated |
| No data | MUC, JBA Kedah, SAJH and SAMB | Data not available or not given. No reasons were given |

[1] Based on the numbers of failed-tested samples against the numbers of tested samples conducted in one year.

# Summary

One of the measures that can help developing countries in meeting Target 10 of the Millennium Development Goals – halving the number of people without access to water and adequate sanitation by 2015 – is through a water sector reform. In this research the Malaysian water sector reform is assessed by answering the following questions:

- How can we understand and explain the policy process of the reform?
- To what extent have the outputs of the reform contributed to the realization of the reform's objectives? and
- To what extent has the water sector reform improved the operational efficiency and environmental effectiveness of water utilities?

This research was approached from two theoretical perspectives: the policy arrangement approach and the policy evaluation approach. The policy arrangement approach provides the analytical tools to ascertain answers to the first question. This is done by thoroughly investigating the main discourses underpinning the water sector reform, the resources-power nexus, the actors and the rules applied and created in the reform process. The policy evaluation approach answers the remaining two questions from two aspects. First, by assessing the output efficacy of four institutional outputs against the reform objectives. Second, by assessing seven quantitative indicators related to the performance of water utilities on operational efficiency and environmental effectiveness.

Chapter 1 takes us through the historical development of the Malaysian water supply sector from as early as the 19<sup>th</sup> century. The first water supply development has taken place in the Federated Malay States and the Strait Settlement under the British administration. Over the years, continuous investments have put Malaysia amongst countries with high access to drinking water in the world. As the country developed and economic development accelerated, public funding proved no longer able to satisfy budgets needed for water infrastructure development. Turning to private sector assistance has neither solved this problem, nor relieved state governments from financial burdens. Political and socio-economic reasons prohibit water from being priced at its actual costs. Low tariffs (and subsidised tariffs) do not only hamper water conservation, but also deprive water utilities from generating extra revenues to sustain operation and to expand services to new areas.

Chapter 2 consists of two parts. The first part presents a general overview of water supply reform processes and explains the key concepts – equilibrium levels, efficiency, effectiveness, equity, competition and unbundling – which are highly relevant to the analysis of water supply reform in Malaysia and beyond. The second part discusses two theoretical frameworks used in this research: (1) the policy arrangement approach for understanding and analysing the reform process; and (2) the policy evaluation approach for measuring the outcomes of the reform on seven indicators: four for operational efficiency – non-revenue water, collection efficiency, unit production cost and customer complaints – and three for environmental efficiency – sludge management, water quality and information disclosure. The last section of this chapter formulates a theoretical framework for this research and explains how this framework helps to answer the research questions.

Also Chapter 3 is divided into two parts. The first part reflects on the leading events of and driving forces behind the reform process. This includes re-visiting the policy decision of the National Water Resources Council made in 2003, the formation of the Ministry of Energy, Water and Communications right after the 2004 general elections, the commissioning of a major study and the amendment to the Federal Constitution. The reform accelerated the desire to address the presence of high non-revenue water, low water tariffs, weak regulation and the unviable financial mechanism for coping with the issue of under-investment and over-dependency on public funding in the water sector. The second part analyses the outcomes of the reform from four perspectives: regulation, water resource management, financial and operational issues. The analysis produced mixed results. Considerable positive achievements are clearly visible with regard to the role of the central regulator on regulation oversight, the financial mechanism of the PAAB and the corporatisation of state water departments. But the state water resources regulator has not shown promising results in safeguarding water resources.

In Chapter 4, the water supply sector reform is analysed using the policy arrangement approach. The analysis shows a shift in water management towards the public control both at the federal and state government levels. This can be seen from the dominating corporatisation discourse, showing a growing tendency towards centralisation of water management. The shared-responsibility approach gives the federal government powers on water regulation and financing matters while reinforcing the dominant role of state governments on water provisioning and water resources. The dominant role of state actors in water is aided by their possession of vast resources – legal, water rights, financial – thus undermining the presence of market actors (private water utilities) in the sector. By the same token, several unprecedented rules emerged from the reform. The most noticeable one is the participation of the Prime Minister and his Deputy during the reform process. Others include the declassification of the public documents, drafts of WSIA and SPANA, from the official secret law to solicit public feedback, and the public hearings process involving members from the opposition parties as well as civil society.

Chapter 5 presents the assessment of four operational efficiency indicators: non-revenue water, collection efficiency, unit production costs and customer complaints. The assessment was furthered investigated in two cases studies – PBAPP and SADA. It became evident that private water utilities are superior in managing water losses and collection efficiency compared to their counterparts in the public sector, while there is no clear distinction with regard to their capability in managing costs and customer complaints. This research re-affirms the contribution of non-reform factors to the performance of water utilities before and after the reform, and between private and public water utilities. The findings from the two cases studies confirm the general findings of this research. This chapter also highlights the importance of performance indicators (as a regulation tool) in the water sector in which successful reform depends considerably on the presence of reliable information.

Chapter 6 analyses the performance of water utilities on three environmental effectiveness indicators: sludge management, water quality and information disclosure. Similar to Chapter 5, private water utilities demonstrate a higher compliance level to sludge management under the Environmental Quality Act 1974 and information disclosure requirements than their counterparts in the public sector. Both private and public utilities have a good compliance level in relation to the National Guideline for Drinking Water Quality Standard 2001 for water quality. Findings from the two case studies confirm the presence of a weak relation between the reform and the performance of water utilities on environmental effectiveness indicators. This chapter highlights the call from water utilities to the government to establish sludge treatment companies to spearhead the research and development activities in sludge recycling and to avoid direct discharge of raw sludge into streams. The majority of water utilities are aware of the mandatory information requirement under Section 29 WSIA. By the same token, civil society groups such as CAWP call for a greater information disclosure transparency by making all documents related to the function of NWSC and water utilities available to the public. Lastly, this chapter proposes for adoption of environmental indicators in measuring the performance of water utilities as well as to strengthen the link between information holders/disclosers and information users to facilitate greater transparency and democratic practices in the water sector.

In conclusion, Chapter 7 answers the research questions as follows: (1) the policy process of the Malaysian water sector reform represents the current global trend in centralising water management within the public domain with a clear division of tasks between policy formulation, regulation oversight and service provision. State actors – the federal government on policy formulation, regulation and financing; state governments (through state water companies) on water provisioning and water resources – become dominant players in the water sector; it reduces the role of private water utilities to only a fraction of activities (i.e. treatment) within the entire value chain of water; and it strengthens close regulation oversight from the regulator. Lastly, there is a growing influence of civil society groups on the water sector; (2) the output effectiveness analysis (of the reform) has produced mixed results. Despite the fact that the central regulator, water corporations and the financier are proven to be the important institutions that meet the objectives, the state water resources regulator has not (yet) shown to be a significant institution in meeting the objective of safeguarding water resources; and (3) due to the timing of this research and limited data availability, it proved difficult to link the performance of water utilities on operational efficiency and environmental effectiveness to the reform.

This study further concluded that the relevance of the policy arrangement approach and the policy evaluation approach frameworks for policy evaluation research are enhanced when they are used in combination. Moreover, the application of the latter to assess water sector performance in the data-poor environments experienced in this research presses for more use of expert judgments to complement incomplete and unreliable quantifiable datasets. It is suggested that further research be carried out over a longer time interval when required data are available, laws are fully enforced and the reform institutions are well functioning. Such research should involve a wider selection of case studies of the entire domain of water utilities using both theoretical frameworks.

The immediate policy recommendations include the call to the government to consider measures to facilitate greater public participation in the policy making process of the reform, the consolidation of water management under a single water body and the establishment of a national and disclosed data bank for the water sector.

# About the author

Ching Thoo Kim was born on 4 March 1966 in the state of Kedah, north Malaysia. In 1986, he entered the National University of Malaysia (UKM) and graduated with a Bachelor's Degree in Mass Communication four years later. After graduation, he worked with several private companies before taking up a career in the Malaysian public service in 1995. In 2003, he graduated with a Master's Degree in Communication Technology from the Universiti Putra Malaysia (UPM). In February 2008, the author joined Wageningen University for his PhD study. His studies at the UPM and Wageningen University were sponsored by the government of Malaysia. Upon completion of his PhD (in October 2012) he took up a post as the Under Secretary of the Biofuel Division with the Malaysian Ministry of Primary Industries and Commodities in the new federal government administrative centre at Putrajaya. The author is married to Duangrat Visai with whom he has two children, a son aged 16 and a daugther aged 11.

Printed in the United States
by Baker & Taylor Publisher Services